This book is dedicated to my late bestie and foragin' pup, Ashlee,
my life partner, and my loved ones who have encouraged and
supported me in all my wild endeavors. Love y'all big.

Quarto.com

© 2025 Quarto Publishing Group USA Inc.
Text © 2025 Whitney Johnson

First Published in 2025 by Cool Springs Press, an imprint of The Quarto Group,
100 Cummings Center, Suite 265-D, Beverly, MA 01915, USA
T (978) 282-9590 F (978) 283-2742

All rights reserved. No part of this book may be reproduced in any form without written permission of the copyright owners. All images in this book have been reproduced with the knowledge and prior consent of the artists concerned, and no responsibility is accepted by producer, publisher, or printer for any infringement of copyright or otherwise, arising from the contents of this publication. Every effort has been made to ensure that credits accurately comply with the information supplied. We apologize for any inaccuracies that may have occurred and will resolve inaccurate or missing information in a subsequent reprinting of the book.

Cool Springs Press titles are also available at discount for retail, wholesale, promotional, and bulk purchase. For details, contact the Special Sales Manager by email at specialsales@quarto.com or by mail at The Quarto Group, Attn: Special Sales Manager, 100 Cummings Center, Suite 265-D, Beverly, MA 01915, USA.

29 28 27 26 25   1 2 3 4 5

ISBN: 978-0-7603-9275-1

Digital edition published in 2025
eISBN: 978-0-7603-9276-8

Library of Congress Cataloging-in-Publication Data is available.

Design and Page Layout: Morgane Leoni
Photography: Whitney Johnson, except Shutterstock, pages 21, 35 (top left & bottom right), 37 (bottom right), 39, 42, 58, 70, 94, 107–108, 111 (right), 114, 120, 131 (right), 136–137, 140–141, 154, 161 (right), 163, 166–169, 181–182; and Alamy, page 78
Illustration: Esté Hupp

Printed in China

GO FORTH AND FORAGE

A Guide to Foraging Over 50 of the
Most Common Edible & Medicinal
North American Mushrooms

Whitney Johnson

COOL
SPRINGS
PRESS

CONTENTS

Introduction: Let's Go Forth and Forage! 6

PART 1—MUSHROOMING ESSENTIALS

Chapter 1: Gettin' Started 9
Regional Field Guides 9
   Latin Binomials 10
Mushroom Groups and Identification Tips 11
Know Your Trees 12
Keep Your Head on a Swivel 13
Mushroom Journaling 14
Take 'Em Home 14
Spore Prints 15
Where to Forage 16
What to Wear and Pack 17
Critters and Bitin' Plants 19
Responsible and Ethical Foraging 21
Ain't Never Not Foragin' 21

Chapter 2: Know Your Parts (Mushroom Anatomy) 23
Cap Features 25
   Warts 25
   Scales 26
   Striations 26
   Zonations 27
Spores and Spore-Producing Surfaces 28
   Gills 28
      True vs. False Gills 30
      Bruising 31
      Is It Milky? 31
      Gill Spacing 32
   Pores 33
   Teeth 33
   Morels and Other Misfits 34
Stipe/Stalk/Stem 37
   Universal/Partial Veil 37
   Skirt/Ring/Annulus 37
   Reticulation 38
   Volva 38
   Basal Bulb 38
Mycelium 39

Chapter 3: Common Mushroom Myths 41
Don't Touch 'Em! 41
Gills Kill! 43
You're Hurting the Mushroom Population! 43
Cut! Don't Pluck! 44
How Can You Tell What's Poisonous and What's Edible? 45

Chapter 4: Words of Wisdom 47
Start Small and Not Raw 47
Do NOT Get Discouraged 48
Be Good to Your Momma (Nature) 48
Ya Ain't Gotta Be Fancy 49
Don't Put That In Your Mouth! 50
Have a Big Ol' Time! 50

PART 2—EDIBLE APPALACHIAN MUSHROOMS BY THE SEASON

Chapter 5: Edible Mushrooms of the Spring 55
Morels (Genus *Morchella*) 56
   Common Morel, Yellow Morel (*Morchella americana*, Formerly *M. esculenta*) 58
      Fried Morels 59
   Eastern Half-Free Morel (*Morchella punctipes*) 60
Pheasant Back (*Cerioporus squamosus*) 62
Wine Cap (*Stropharia rugosoannulata*) 64
Wood Ear (*Auricularia angiospermarum*) 66
Deer Mushroom (*Pluteus cervinus*) 68

Chapter 6: Edible Mushrooms of the Summer 71
Chanterelles (Genus *Cantharellus*) 72
   Golden Chanterelle (*Cantharellus* sp.) 73
   Smooth Chanterelle (*Cantharellus lateritius*) 77
   Peach Chanterelle (*Cantharellus persicinus*) 77

Appalachian Chanterelle (*Cantharellus appalachiensis*)	78
Cinnabar Chanterelle (*Cantherellus cinnabarinus*)	80
Black Trumpets (*Craterellus fallax*)	82
White and Black Trumpet Pizza	84
Chicken of the Woods (Genus *Laetiporus*)	85
Chicken of the Woods, Yellow (*Laetiporus sulphureus*)	86
Chicken of the Woods All-Purpose Seasoning Salt	86
Chicken of the Woods, White (*Laetiporus cincinnatus*)	88
Tawny Milk Cap (*Lactifluus volemus*)	90
Puffballs	92
Giant Puffball (*Calvatia gigantea*)	93
Grilled Puffball Steaks	95
Common Puffball (*Lycoperdon perlatum*)	96
Brain Puffball (*Calvatia craniiformis*)	98
Pear-Shaped Puffball (*Apioperdon pyriforme*)	99
Beefsteak Polypore (*Fistulina americana*)	100
Old Man of the Woods (*Streobilomyces floccopus*)	102
Shaggy Mane (*Coprinus comatus*)	104
Reishi Mushrooms (Genus *Ganoderma*)	106
Hemlock Reishi (*Ganoderma tsugae*)	107
Reishi Tea	109
Berkeley's Polypore (*Bondarzewia berkeleyi*)	110
Berkeley Jerky	110
Indigo Milk Cap (*Lactarius indigo*)	112
Oyster Mushrooms (Genus *Pleurotus*)	114
Summer Oyster (*Pleurotus pulmonarius*)	115
Wild Summer Mushroom Toast	117
Rooted Agaric (*Hymenopellis furfuracea*)	118
Boletes	120
Frost's Bolete (*Exsudoporus frostii*)	121
Crown-Tipped Coral (*Artomyces pyxidatus*)	123

Chapter 7: Edible Mushrooms of the Fall
	127
Oysters (*Pluerotus ostreatus*)	128
Cream of Wild Mushroom Soup	130
Hen of the Woods (*Grifola frondosa*)	131

Hen of the Woods Flatbread	133
Lion's Mane (*Hericium erinaceus*)	134
Lion's Mane "Crab" Cakes	133
Bear's Head Tooth (*Hericium americanum*)	136
Coral Tooth Fungus (*Hericium coralloides*)	138
Wood Blewit (*Collybia nuda*)	140
Honey Mushrooms	142
Bulbous Honey Fungus (*Armillaria gallica*)	143
Shrimp of the Woods (*Entoloma abortivum*)	145
Popcorn Shrimp of the Woods	147
Amber Jelly Roll (*Exidia recisa*)	148
Amber Jelly Roll Gummy Candies	148
Snow Fungus (*Tremella fuciformis*)	150
Cauliflower Mushroom (*Sparassis americana*)	152
Lobster Mushroom (*Hypomyces lactifluorum*)	154
Lobster Mushroom Duxelles	154
Resinous Polypore (*Ischnoderma resinosum*)	156
Purple-Gilled Laccaria (*Laccaria ochropurpurea*)	158
Wrinkled Cortinarius Mushroom (*Cortinurius caperatus*)	160
Shaggy Stalked Bolete (*Austroboletus betula*)	162

Chapter 8: Edible Mushrooms of the Winter
	165
Late Fall Oysters (*Sarcomyxa serotinus*)	166
Chaga, Medicinal (*Inonotus obliquus*)	168
Enoki (*Flammulina filiformis*)	170
Yellowfoot (*Craterellus tubaeformis*)	172
Wild Yellowfoot Quiche	173
Hedgehog Mushroom (*Hydnum* sp.)	175
Hedgehog Mushroom Bacon Penne Pasta	176
Turkey Tail, Medicinal (*Trametes versicolor*)	178
Double Extraction Turkey Tail Tincture	180
Witches' Butter (*Tremella mesenterica*)	181

Conclusion	183
Resources	184
About the Author	185
Index	186

INTRODUCTION

Let's Go Forth and Forage!

Howdy, foragin' friend! Whitney here, or you might know me by my social media handle, Appalachian Forager. Whatever ya call me, I am just tickled that you're here and have chosen to place this book into your paws. I might be a little biased, but I think you've made a good choice if you are a baby mushroom hunter and need an easy guide to expand your mushroom huntin' knowledge and experience in the hills and hollers.

My love for the outdoors developed at a young age, and as far back as I can remember, I've been hard to keep in the house. As a wee holler baby, you could find me outside digging in the dirt with spoons (sorry for ruining all the eating utensils in the trailer, Mom). I was a professional mud pie maker, bass fisher, snake wrangler, and tree climber by the age of 5. I reckon it was only a matter of time before my love of the outdoors and food would collide and open my eyes to the best hobby there is in my opinion, mushroom huntin'!

My devout love for the fungi came a little bit later in life when I was in college, around my sophomore year. I was not my healthiest self at that time, mentally or physically. I was in a dead-end relationship, kinda down in the dumps, and I was yearning for something more. I made it my mission to walk with Mother Nature as often as possible and get out into the woods to heal my soul. Turns out, it works wonders! Not far into this mission, I started to notice something I hadn't paid much attention to before—mushrooms. They were all over the trails I was traipsing. There were just *so* many! These mushrooms were of all colors, textures, and sizes, and I was enamored right off the rip.

An absolutely chonky common morel mushroom (*Morchella americana*).

I could not believe that I had not taken in their beauty and abundance until that season of my life. I'm into snappin' pictures, too, so I started taking photos of every single mushroom I came across. These photos sparked my curiosity, and I wanted to figure out what kind they were and whether or not I could eat them, of course. I was a poor college gal and couldn't afford to buy field guides at that time. I found an online mushroom guide and printed off all 600-plus pages in color from the printer at my workplace. I almost got fired for that when my boss found out. Anyhoo, I put those printed pages into a three-ring binder and toted it with me on my woodland adventures. I imagine I probably looked a little bit ridiculous sitting on my hind-end trailside comparing random mushrooms to my rickety binder pages, but that's okay. I was happy and I was learning. That all snowballed, and I went headfirst into the mushroom foraging world. I haven't stopped or looked back since, and what a beautiful whirlwind journey it has been, honey.

Fast forward a few years, and this gal from the boonies of Blaine, Kentucky, has gotten some experience under her belt, and I have learned all that I possibly can about this region's fungi offerings. I have logged a metric butt-ton of woman hours and put thousands of miles on these here feet in search of mushrooms. I now consider myself to be a seasoned, self-taught mountain mycologist preachin' the good word to the masses.

Foraging has helped me in more ways than I can even express on these here book pages. It has given me an extremely fulfilling, satisfying hobby, while allowing me to reconnect with my Appalachian roots and with nature—the best medicine there is. Taking foraging to my social media platforms has afforded me the opportunity to shine a loving light on my region and to educate others on what wild food is out there for grabs, and now here it is in book form! My social media following has continued to grow, and it tickles me to death to know there are so many others out there who share a passion for mushroom huntin'.

So, what separates this guidebook from the rest of the mushroom field guides that are out there? I'll tell ya what. I have created this guide for *you*. Yes, I mean you. This is an absolute bare-bones beginner's guide to mushroom hunting for anyone and everyone. Some guides can be a little intimidating and showy, and that is exactly what this one ain't. This book is written in a down-to-earth language and style that is easy to understand. Yeah, there may be some lingo and terminology you might not be familiar with if you're just getting your start, but my vow to you is that I will explain it in a way that is foolproof. I'll cover all the major bases and get you pumped up to get out there and explore.

GO FORTH AND FORAGE!

Findin' a creekbank full of smooth chanterelles never gets old and will make ya holler every time.

PART 1— MUSHROOMING ESSENTIALS

CHAPTER 1

Gettin' Started

Starting new things is exciting, but it can sometimes also be scary. You can feel quite intimidated when you decide to jump into a new-to-you thing, and that's just fine, dandy, and totally normal. I mean, you'd be kinda crazy if you didn't have at least a jag bit of some anxiety knowing that if you ate the wrong mushroom, it could cause your early expiration. But don't let that freak ya out because here I am, comin' to the rescue!

In this chapter, I will give you all the information that I sure wish someone had given me when I was getting' my fungi foragin' start. I'm gonna cover things from top to bottom to get you comfortable and to calm those nerves as it relates to mushroom huntin'. Ya ready?!

REGIONAL FIELD GUIDES

When anyone asks me for the first step in starting their mushroom hunting journey, my suggestion is always gonna be to get a regional field guide. And butter my backside and call me a biscuit! Ya got one in your hands right now. Step one complete! The guide you hold currently contains information on some of the edible species that are around you, so if you are interested in knowing other edible and inedible mushrooms, you can opt for a larger field guide down the road.

Field guides are purdy much like your mushroom bible. They serve as a comprehensive guide for the mushrooms that occur in nature. In these guides, you'll find descriptions and pictures to help you better understand the fungi around you. These guides will include a ton of information on the mushrooms, including details about their caps, flesh, spore-producing surfaces, stem, habitat, spore print color, distribution, and sometimes edibility. But not all guidebooks are created equal. There are guides that encompass the entire North American continent, and these guides can be daunting for someone who is just getting started. If you were to start with a field guide this large, you may cause yourself some confusion with a book that includes mushrooms that do not even grow around you. Some field guides can be lengthy, complicated, and not the best choice for beginners. The good news is that there are easier, smaller books to utilize initially. This is why I suggest beginning with *regional* field guides. *Regional* means that the book pertains to your area that you live and breathe in.

When you crack open your first honkin' field guide, you may start reading and think, "Oh, law! I'm not smart enough for this!" Well, listen here, honey, I've been there and done that. I bought the t-shirt. They can be intimidating at first, but I promise that if you are interested in learning, it will come easier and more naturally than you might think. As long as you keep at it, you'll be navigating these books (including this one!) like a pro and understanding all the different chapters and mushroom terminology contained within.

LEFT | A fistful of golden summer chanterelles (*Cantharellus* sp.)

Latin Binomials

In this and every other mushroom field guide, at the tippy top of the page, each mushroom will have a name that is referred to as its *Latin binomial*. Latin binomials are used to identify different species throughout the animal, plant, and fungi kingdoms. *Binomial* means that they are broken down into two names: "bi" meaning two and "nomial" meaning name. The reason we use Latin binomials is because it serves as a universal language for us to identify and discuss different mushrooms across the world. In addition to the mushroom's Latin binomial, guides will typically include an alternative "common name." Common names are a much simpler name for the mushroom that isn't a foreign mouthful to attempt to pronounce.

Here's an example of why Latin binomial names are important. There is an edible mushroom with the common name "chicken of the woods." Now, there are actually a few different types of chicken of the woods, and each type of chicken of the woods has its very own Latin binomial. If I called up a buddy and said, "Hey, sis! I found chicken of the woods!" they wouldn't know exactly what mushroom I was talking about. The two most commonly occurring chicken of the woods in my area go by the Latin binomials of *Laetiporus sulphereus* and *Laetiporus cincinnatus*. *Laetiporus sulphereus* has a brighter orange cap with yellow margins, a porous yellow underside, and is usually found growing in shelves directly on dead/dying hardwood trees. *Laetiporus cincinnatus* has a more pinkish or peachy cap, a white porous underside, and is commonly found growing in rosettes from buried, dead wood, appearing to grow from the ground. So, you can see that even though they share the same common name, they are two different mushrooms with different features. It's important to understand the Latin binomial so we can be certain what mushroom we are referring to. With chicken of the woods, it is fairly normal for some folks to get a sick belly when eating *Laetiporus sulpherus*, but they can better tolerate and enjoy *Laetiporus cincinnatus*. Knowing the Latin binomial of the mushroom eliminates all confusion and ensures you're not eating something you don't want to.

Another reason Latin binomials are important is because, depending on your region, you may refer to a mushroom by a different common name than someone from another part of the world. Morels come to mind. There are more common names for morels than Carter has liver pills. Someone from the other side of the country may have no idea what "dry land fish" are. If you confirm the morel you're talking about with the Latin binomial, *Morchella americana*, then we're all pickin' up what you're puttin' down.

Fret not, lil' forager! It is not mandatory to learn and memorize all the Latin binomials of mushrooms. It is not necessary in order to go out and mushroom hunt, but it is beneficial to understand the concept and to know when you may need that information. You can always access Latin binomials via your field guide when you are working to identify your specimen or when you're needing the Latin binomial to discuss mushrooms with fellow mycophiles.

MUSHROOM GROUPS AND IDENTIFICATION TIPS

Another great resource to utilize is mushroom identification (ID) groups. There are many of these communities available to you on Facebook. They're places where folks can submit photos of mushrooms they have found when they aren't quite sure what they have. It's perfectly okay to find a mushroom and be conflicted, especially when you're just cuttin' your teeth. If you have scoured your field guide and have any apprehension at all, mushroom ID groups can help! Members of the ID group will comment and help the poster in identifying the particular mushroom through pictures and information the poster shares. I used these groups many times when I was getting started and even will refer to them when I'm stumped today. The responses are quick and accurate.

Within these groups, there are steps to take and features to include in photos to get proper identification. These rules are great to apply across the board when asking for assistance with mushroom identification. So, if you wanted to get help with figuring out a mushroom, whether it be asking another mushroom buddy, posting in one of the identification groups, or even if you're shooting me a DM to help you identify, you want to include all of the following things:

First off, ya want angles, baby. When you see a mushroom, make sure you photograph the top, the side, and the underside of the mushroom in its natural growing habitat. Photograph before you pluck it! You also want a photo to include the stem (if present). Make sure the photos are clear and not blurry. If I had a nickel for every time I received a mushroom ID request and it was a blurry photo of only the top of the mushroom, I'd be a daggone millionaire. By getting all the mushroom's angles, it ensures all the features of the mushroom are included, which helps for a more solid ID.

Here's a right good example of picture angles to get when requestin' mushroom identification. You want to be able to see the top, side, spore producing surface, and stem. Make sure to include other pertinent information such as location and habitat.
*This is an edible meadow mushroom (Agaricus campestris)

After you've gotten these angles in photos, you then should carefully excavate the entire mushroom, including the base. The base of a mushroom can be different sizes, shapes, and so on, and can serve as a key identification feature, making this an important step.

Next, and I know this sounds weird, make note of any smell the mushroom may have. I sniff every single mushroom I pick. Some mushrooms have very distinct smells. For example, wild enoki mushrooms have a strong scent of rust, and that's a good way to help figure out that it *is* enoki and not a lookalike that doesn't have this smell to it.

Be sure to document any trees the mushroom may be growing around. This is very crucial information because certain mushrooms love to grow near certain types of trees.

Lastly, make sure you include information about where you found it. A lot of mushroom identification groups are international, so advise what state and country you found the mushroom in, as well as the date/time of year since many mushrooms are seasonal and regional. Once you check all these boxes, you're well on your way to finding out what mushroom ya got.

Gettin' Started

This big, beautiful hemlock tree was just begging for a bear hug. She is one of my favorite cold-weather mushroom trees.

KNOW YOUR TREES

I cannot stress this one enough. Get. To. Know. Your. Trees. Certain mushrooms love certain trees and are known to occur in each other's company. This is what we call a *mycorrhizal relationship*. The tree and the mushroom help each other out by combining their mighty powers. Tree roots and mushroom mycelium (the rootlike system of fungi) will hook up and swap water and nutrients to help all parties involved grow big and strong. Ain't nature awful sweet?

Once you become privier to your trees and which mushrooms tend to have relationships with those trees, it's gonna make finding that mushroom a heck of a whole lot easier. For example, in the picture to the left, you will see me giving a big ol' bear hug to a hemlock tree. My brain that is now hard-wired for mushroom hunting sees that hemlock tree and immediately starts thinking, "What mushrooms grow around this feller?" I now know I can find hedgehog mushrooms, yellowfoot, reishi, and certain types of boletes and milk cap mushrooms around that type of tree. Another benefit of knowing your trees is, if you are targeting a specific kind of mushroom, you can put yourself in a forest containing the trees your mushroom likes to hang out with. If you plan on hunting down some chanterelles in the summer, you'll know to look for a forest rich in beech, oak, poplar, and birch trees.

Many guides often distinguish whether the mushrooms fruit with hardwood/deciduous trees or conifer trees. Most hardwood/deciduous trees are trees with broad leaves that produce a fruit or nut, shed their leaves in the fall, and go dormant in the winter. Conifer trees are cone-bearing trees that are typically evergreens, meaning they keep their needlelike foliage year-round.

Just as there are mushroom field guides, there are also tree guides. If you are interested in beefin' up your knowledge of the woods, you can consider adding one of those to your collection.

Once you've learned your trees, you've won half the battle. Seriously.

TREE TYPES

Here's a quick guide to go by when you're navigating this book and any others that list the types of trees a mushroom is found growing with:

Common Hardwood/Deciduous Trees:

ASH	SYCAMORE	POPLAR
WALNUT	BIRCH	CHERRY
ELM	BEECH	ASPEN
OAK	ALDER	
MAPLE	HICKORY	

Common Conifer Trees:

PINE
HEMLOCK
CYPRESS
CEDAR

KEEP YOUR HEAD ON A SWIVEL

Foraging is a 360-degree, panoramic experience. My main advice to you when you start hitting the hills looking for wild mushrooms is to quite simply, LOOK EVERYWHERE. Mushrooms are everywhere. They grow from the ground, behind bark, on dead trees, on live trees, near water, in your yard, next to the sidewalk, on fence posts, from flowerbeds, or even from toilets. I mean it. I've had people ask for my help in identifying their toilet mushrooms, but I usually just tell them to clean up their bathroom a little better.

One of the biggest mistakes that can be made when mushroom hunting is solely focusing on one area of your surroundings. By this I mean that you might find yourself only looking at the ground. And let me tell ya, you're doing yourself a major disservice if so. If you have your nose to the ground like a bloodhound, you might miss out on some top-choice edible mushrooms growing above on trees. It is very easy to walk right past a tree mushroom if you're not paying attention. I've done it before! Mushroom hunting is definitely not a fast-paced sport. If you're going mushroom hunting, you should expect many stops, some off-trail moments, and a slow, steady stride. You want to give yourself the time and opportunity to soak in the territory and scan the land for goodies. Look up, down, and sideways, because mushrooms sure love to grow all kinds of cattywampus, and you don't wanna miss 'em.

MUSHROOM JOURNALING

Something that helped me immensely when I was a baby mushroom hunter was creating my very own mushroom journal. This is a way for you to log your finds and become more confident with your catalog of mastered mushrooms. It's also a great reference to flip back through to see your progress and learn fungal patterns.

You could opt to do your mushroom journaling in whatever style that fits your vibe. You could literally have a journal that you put pen to paper on, you could take notes in your phone, or you could have an online blog or space to record information—it's really whatever you prefer.

Pictures are, in my opinion, the most important thing to include in your mushroom journal. Having your own personal picture that *you* took will be an excellent reference to have along with any other information you've gathered. A personal photograph is so much better than a stock photo you may see in another book or online because you were actually there! You took it! It will allow you to see the mushroom and look at its special features and its unique appearance.

Location is another important key piece of information to include. It's nice to be able to see what park, state, or area you found the fungi in. You could even go as far as to include latitude and longitude if you wanted. Why is this important? This will give you the ability to see where you found certain mushrooms and can clue you in to look in surrounding areas for those same mushrooms.

The date your mushroom was found is also crucial to note. This will provide seasonal information for you to familiarize yourself with around when that certain mushroom tends to grow. You can check your journal and realize you found that mushroom on that date in that spot the prior year, which means you should probably go check again at the same time this year. For example, the cauliflower mushroom is a mushroom that often will fruit in the same spot around the same time of the year. If you check your journal and realize you found that mushroom this time last year, you can go score it again potentially from that very spot.

Additionally, note any surrounding terrain around the mushroom. This includes trees, soil, water, and so forth. It will help you to understand what habitats that mushroom prefers, strengthening your skills to get yourself in the right place for target mushrooms.

Include weather conditions. You can add in if it was rainy, cold, hot, dry, snowy, and so on. In doing so, you'll learn what kind of weather yields which mushrooms.

It's always nice to write into your journal what you did with that mushroom. You could add in whether or not you took it home, compared it to field guides, did spore prints, submitted it to ID groups for designation, or whatever. The more information there is for you to look back on and see your growth and experience, the better.

And finally, you have got to include those recipes! If you brought home a tasty, edible mushroom and made a bangin' recipe with it, don't let that slip through your fingers. I have been guilty of cookin' up some dang good dishes and not even writing down how I did it. You're gonna wanna have these for your records in order to duplicate a recipe later if it was a hit.

The points I've listed here are what I personally add into my journal, but you can personalize this to your liking. Add or omit as you wish to best document your glow up from a baby forager to a confident, unstoppable mushroom hunter.

TAKE 'EM HOME

You might think the only mushrooms you want to bring into your humble abode would be the ones you are for sure gonna eat, and if that's what you feel comfortable doing, that is fine. However, I know that bringing mushrooms I am not sure of home to further investigate helped me a whole lot. I touched on this in mushroom journaling, but let's go a tad deeper, shall we? Do *not* be afraid to take them babies home! I mean this for all mushrooms. Yes, even the ones that may not be edible. When you are out on the trail and looking at mushrooms, you may not have ample time to sit there and figure out what you're looking at. You may not have cell phone service to check in with someone to help you ID. By throwin' that puppy into a sack and bringing it home, it's going to give you more time and space to research the mushroom. I do recommend keeping questionable mushrooms separate from your edible mushroom sack, just so we don't mix anything funky with the mushrooms we will be chowin' down on later.

Once you're home with the mushroom you want to do more work on, the world is your oyster (mushroom). Heh. You can then leisurely compare it to your field guides and partake in some Internet research and have that mushroom right next to ya. Sometimes, the pictures you took in the field just don't do it justice when you're doing further research later. Work smarter, not harder. Having it there in real life will let you engage all your shroomy senses. You have the information right at your fingertips of what it smells like, feels like, what the cross section looks like, and so forth. You can also use this information to submit to mushroom ID groups if you are still unsure.

Once you have nailed it down and ya know what mushroom ya got, then you can decide what you'd like to do with it. If it's edible, you may want to cook it up or preserve it for later use. If you find you have a toxic or inedible mushroom, don't panic! You can discard it outside, away from any areas where pets or children may roam. Make sure to clean up any bits and pieces you may have sprinkled about in your home during your quest for information.

SPORE PRINTS

If ya ain't done a spore print, ya ain't lived. This is one of my favorite things to do with mushrooms for both ID purposes and because it's artsy and fun.

The deer mushroom *Pluteus cervinus* yields a salmon-pink spore print. This one I had sprayed to preserve as wall decor, so the color changed somewhat, but you can see the definition of the gills so perfectly.

Some mushrooms will throw a white spore print. If I'm expecting a light-colored spore print, I like to use a dark-colored paper to collect it, like with this broad gill (*Megacollybia rodmani*) I was testing out.

Gettin' Started

You'll feel like a mad scientist in your laboratory when doing them. A spore print is a dusty deposit the spore-producing surface of the mushroom will drop onto a surface after it has sat for a spell. Different mushrooms leave different color spore prints. Spore prints can be all the colors of the rainbow. The color of the print is significant and will help you confirm or reject your ID hypotheses. Allow me to share an example of where a spore print would really come in handy. In the wintertime, you may be on the hunt for edible enoki mushrooms (*Flammulina filiformis*). Enoki has a toxic lookalike called the funeral bell or deadly gallerina (*Galerina marginata*). The two do have some similar looking features, but a sure way to separate them is by the color of their spore prints. Enoki mushrooms have a white spore sprint, whereas funeral bells have a rusty brown spore print.

So, how do you even *do* a spore print? It's so easy, bubbies and sissies. All you need is a mushroom, a flat surface, something for the spores to fall onto, like construction paper or aluminum foil, and a glass or bowl to cover the mushroom. What ya do is lay down your paper or foil onto a flat surface. You then put the mushroom spore-producing–side down onto the paper or foil. This means gill-side down, pore-side down, tooth-side down, or with club fungi or morels, you just lay it down. If the mushroom has a large stem, I snip it, so that way I am able to put the cap down directly onto the paper or foil. Some folks like to put a droplet of water on the top of the mushroom cap to encourage it to drop spores, but typically, I don't have to do this step. If you have a healthy specimen, it should be just fine. Next, cover it with the glass or bowl to protect it while it does its thang. Leave it to brew overnight. Come back the next day and pick up the bowl or glass and flip up the mushroom. Voila! It has left its cute lil' spore print. You can then observe what color it is and whether or not the color lines up with what you're after.

Fun sidenote, I like to do spore prints and spray them with a fixative spray and frame 'em for art. I've used Aqua Net hairspray, and it works real nicelike. You know if it held those hairdos in the 1980s, it will surely hold spores to a piece of paper.

WHERE TO FORAGE

I get this question often: "Where can I go to forage mushrooms?" There really isn't a blanket answer for this since it depends on your location. Different state and local governments have their own rules that vary from place to place. My main advice would be just to do your research wherever it is you plan to look for mushrooms. People often think of foraging to be off-limits in national parks, but individual parks have the authority to either allow or not allow certain things to be foraged. You may be surprised to know that 75 percent of national parks allow foraging of some kind!

My primary stomping grounds are in eastern Kentucky. In my state, I can collect mushrooms in National Forests and Wildlife Management Areas with the maximum haul being 1 gallon (4.4 L) of mushrooms per day per person. You must have a permit if you want to gather more than a gallon (4.4 L) or if you are planning to sell said foraged mushrooms. Different states will have rules like this, so please do your research so you're a legal beagle and not a mushroom outlaw on the run.

If you have seen any of my videos, you may think, *well, you get way more than a gallon of mushrooms on your hunts*, and you'd be right. That is because I am foraging on private property where there are typically no limitations or restrictions on how you forage other than what the owner allows. With that being said, *always* get permission to go onto private property. In my neck of the woods, it's very dangerous to be roaming on someone's property unbeknownst to them and could result in your murder. I will always ask the landowner if I can frolic through their acreage before I ever step foot into their territory.

I have no shame in my game, and I do frequent drive-bys. If I'm driving by your house and I see a nice flush of edible mushrooms growing on a tree in your yard, I will knock on your door and ask if I can have some. I'd say that 99 percent of the time, that person has no plans for the mushrooms and will let you take some bounty. If they say no, big whoop. It doesn't hurt to try, and there's still plenty other mushrooms to be had!

If you are left with any doubt about whether you can forage in a specific area or park, call the park office. This is an easy way to find out what the rules and regulations are. You can do internet research, but hearing it straight from the horse's mouth will ease any fears of illegal mushroom hunting activity.

WHAT TO WEAR AND PACK

I'll say it over and over until I'm blue in the face: Ya ain't gotta be fancy. Foraging is a gritty, primal practice, and you don't need all the bells and whistles to do this thing. When it comes to what I'm wearing, it sure ain't any high-dollar brands and is usually something thrifted that gets the job done just fine.

For footwear, I recommend waterproof hiking boots. Mushroom hunting will inevitably have ya in the mud and doing some creek crossings. If you're more of an only-stay-on-the-trail forager, you could get whatever type of boots fit your adventure. Ya just make sure you get yourself something comfortable and with good traction because you wanna be able to scale them hills with ease if you spot a mushroom up yonder.

For clothing, I like to wear something comfortable. Your clothing should not be so tight that it will restrict your movement and also not so baggy that it's getting caught on everything or letting bugs

A simple and light hiking pack works best when mushroom hunting. I fill it with my foraging essentials and ain't weighed down as I lollygag through the woods.

shimmy up onto your bare skin. You'll definitely want to dress dependent on the weather conditions. In the warmer months, you'll usually find me in a tank top and leggings. I never wear capri length pants or shorts while hiking just to keep those ticks off me. I like to also tuck my leggings into my socks to create another barrier for bugs. In the colder months, make sure you're wearing enough layers to keep yourself warm because nobody likes frostbite or cold feet. *Always* bring extra socks with you. I've noticed if I don't bring them, I need them. If you somehow get your socks wet or find your feet getting cold, that spare pair will be a godsend.

So, what to pack? Again, I ain't fancy. My foraging pack was $20. It's an outdoor pack that has multiple pockets, and it has a place to put a CamelBak for water. It also has a nice rain guard you can pull out to cover the pack if you get caught in a downpour. I love having a rain guard feature because I can put my precious cargo of mushrooms into my pack and protect the goods from getting soggy.

In my pack, I will bring water and snacks because you don't want to be thirsty or hangry in the hills. I also bring a sweat rag. I sweat at all times of the year, so it's nice to have something to dab the perspiration before it gets into your eyes. I bring a

A crossbody mesh bag is my go-to for mushroom foraging. It's hands-free, allows my mushrooms to breathe, and lets the spores fall through and sprinkle the forest floor to grow new mushroom babies.

Mushroom knives are designed specifically for harvesting mushrooms and often come with a brush on the end to clean off your goodies in the forest before you add them to your sack.

field guide with me that won't weigh me down too much (this one, for example). Definitely fight the urge to bring your library of field guides in your pack. If you toted them all, you're liable to develop a humpback from lugging those heavy things around the woods. I also pop a simple first aid kit into my pack just in case of accidents. I'm a little accident-prone, so the first aid kit has been clutch for me quite a few times. Mine includes Band-Aids, gauze, an ice pack, antiseptics, and the like.

Ya gotta have something to put your mushrooms in. You could opt for a mushroom bag or a mushroom basket. I use a mushroom bag that is mesh. Mesh is best because while you're skipping through the hollers with your sack full of fungi, the spores can disperse through the holes and drop onto the forest floor to encourage further mushroom babies. My sack is a very simple crossbody bag with a hanging pocket to put any tools in that I need to whip out quickly on a hunt. You *could* use a basket, but I have found this to be more of a romanticized idea that doesn't really line up for my type of forays. I am often off-trail, in the thick of briars and branches, so a basket gets all hung up and it just turns into

a disaster. If you plan to forage in an open field or chiefly on the trail, you use that basket, honey.

Ya gotta bring a handy-dandy mushroom knife. Mushroom knives are specifically designed with us foragers in mind. The blade will be slightly curved for optimum mushroom sliceage, but the real stunner here is the mushroom brush. Most mushroom knives will have a retractable brush on the end to dust any dirt and debris off your fungi. If you have a sack of mushrooms that are fairly clean and you throw in just one dirt-covered mushroom, they are gonna bang and clang around, which results in a bunch of now-dirty mushrooms. I like to take my knife and cut the dirty butts of the stems off and then clean them up with the brush before they make it to my sack. This will keep your mushrooms clean as a whistle, and you'll for sure be thanking yourself in the kitchen later.

Some other things I like to bring would be bug spray, a camera or phone to take pictures, toilet paper, and extra bags for collecting mushrooms I may want to research later. I don't put questionable mushrooms into my main bag with edible mushrooms I'll be munching on for supper.

Now, I'm gonna holler this at you for a second, please consider bringing extra bags to pick up and pack out trash you may come across. Unfortunately, some people do not understand the "leave no trace" rule. I've carried out bags and bags full of garbage. I don't know how many times I've found XL pop cups scattered around. I often wonder what possessed someone to bring an XL pop on a nature hike. It was heavier when they carried it in, why not carry it out now that it's empty? But, I digress. If you see some trash, pick it up. I like to be good to my Momma Nature, and I'd like to think in doing so, she rewards me with a big bunch of mushrooms.

CRITTERS AND BITIN' PLANTS

If I had to choose one thing I've heard people say is the reason they are scared of getting out and foraging, it's the critters, especially snakes and ticks. I hear ya. It is a valid worry, but with the proper awareness and information, you can get out there and put your mind at ease when it comes to the creepy crawlies and slitherers.

Let's start with snakes since they seem to be the primary aversion. I see so many comments on my posts that are like, "Girl, aren't you afraid of snakes?!" The answer is no, but I *am* aware of snakes. In my neck of the woods, we do have some venomous snakes, namely copperheads and timber rattlesnakes. Guess how many I have ran across in all my foraging days? Three. The answer is three. And I have been doing this for a hot minute now, buddies. Each venomous snake encounter was pretty lackluster, thankfully. I've seen two copperheads, and I simply did a wide right turn to get around them and not disturb them and I didn't die. Another time I tangoed with a nope rope was kind of a funny story actually. I was doing a guided hike in the Red River Gorge in Kentucky. I always do a spiel before I take off with a group of how snakes exist, but that we rarely see them, but how to handle it if we do. Then, as we started walking up the trail, I'll be danged if there wasn't a giant timber rattlesnake that looked like it had swallowed an infant. It was huge. It was just basking in the sun, obviously very full and satisfied with whatever its recent meal was. But we just kinda went around it and no one was harmed.

This is not to say you'll never run across a snake; you probably will if you really get out into nature a lot, but there are things to know and do. First off, be respectful of the area because it's not your home, but it *is* theirs. Don't go flipping logs and rocks and poking your hiking stick into random places. Be aware of your surroundings and check places where snakes like to hang out before you stick your hand in any hidey holes. If you do see a snake, leave it alone. If you're not confident in knowing whether or not it is venomous, treat it as if it is. In saying that, I don't mean kill it, but just don't tango with it. Freeze and back away slowly until you're

Look reaaal close and you can see a rattlesnake takin' a rest after what seems to have been a hefty snack. I kept my distance. I didn't bother it, and it didn't bother me.

at least 10 feet (3 m) away. Once at a safe distance, you can go around it or the snake will leave and find its own way.

Next up, the ticks. Shew. I hate ticks, and they seem to get smaller and more abundant every year. It's dang near impossible for me to go out in the summer and not pick up a few ticks. I've tried all kinds of different tricks, and I wanted to share what has worked best for me. Firstly, light-colored clothing will allow you to see ticks on you better than dark clothing. Long pants and sleeves are good to keep the ticks off you if you can stand wearing them. I use a special spray that pretreats my foraging outfits and is good for six to eight washes. This has been a game changer. I also double-down with tick-specific bug spray right before I hit the hills. When you do get home, make sure you check yourself thoroughly. Bonus points if you have a partner that can check every nook and cranny. My partner and I do this and like to wear head lamps and use surgical tweezers to remove any butt-crack ticks. It happens, and that's true love.

Spiders, and especially spider webs, will be all over the dang place while foraging in the warmer times of the year. It's just part of it, y'all. I'll be honest, ya girl was kinda spooked by spiders for most of her life, but I've really come around in my older age to appreciate them. Although my fear of the spider itself has waned, I will still involuntarily contort, yelp, and look like I'm doing my best impression of *The Matrix* movie when I walk into a big ol' web while hiking. One of my favorite tools on the trail is my handy, dandy, and trusty spider stick. What's that, you ask? It's just any stick I can find on the ground with some nice prongs on the end. I will swing it up, down, and all around in front of me as I'm trekking, and the stick catches the webs before my face does. Being the weirdo I am, I actually have some trails where I leave my favorite spider sticks strategically hidden at the beginning so I can use them each time I'm there. I also name the sticks. My favorite spider stick is Mick.

Always check around downed trees and wood before harvesting your mushrooms as this is a hotspot for snakes to chill.

I always place him in a tree at the mouth of one of my favorite chanterelle spots. He is so reliable and ergonomically put together. Get'cha a spider stick, and even name it. You won't regret it.

Of course, you've got your stingin' and bitin' bugs—mosquitos, bees, yellowjackets, wasps, hornets, and the like—and you're liable to run into them at some point. For mosquitos, if they tend to be fond of ya and you're extra sweet like me, spritz on repellent to protect yourself. If you see any nests or hives of the stingy fellers, leave it be and keep yourself and your pets away. If you do accidentally disturb a nest or a swarm of any angry stingers, slowly walk away and cover your face to protect that precious moneymaker. Flailing your arms and throwing a hissy fit will usually attract the swarm even more. If possible, make your way toward some thicker brush and vegetation to throw 'em off.

Depending on where you live, you could also potentially encounter bears, coyotes, bobcats, or other wild beasties. The general rule of thumb here is pretty much the same as we discussed before. You should remain aware of your surroundings, avoid places where certain critters are known to hang out or reside, and always be respectful of the wildlife's home. Most animals will go the other way if they see you. If you are very concerned, you could carry bear mace and tie a bell to your pack so the noise deters critters from the area.

Just like the critters, there are some plants that occur in nature you'll want to watch out for and steer clear of, too. These can include poison ivy, poison sumac, poison oak, Virginia creeper, and wood nettle. See the illustrations below so you can better learn to identify and avoid these plants that can reach out and "bite" ya.

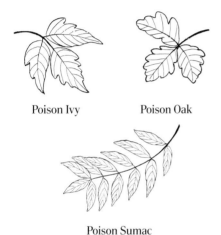

Poison Ivy   Poison Oak

Poison Sumac

All in all, yes, you might see some critters or brush against some ouchy plants at some point, but as long as you're safe, aware, and prepared, you *can* forage without fear!

RESPONSIBLE AND ETHICAL FORAGING

When you finally experience the rush of finding an absolutely chonky, huge flush of mushrooms, you may get excited and want to cram 'em all into your sack. It happens. It's a high that I cannot express with words. If you've experienced it, you get it. However, there are a few things to keep in mind when it comes to ethically harvestin' your score. You want to leave some bounty behind for a few different reasons.

Firstly, if the mushrooms are young pups, they haven't released their spores yet. I like to leave young, small mushrooms behind because they won't add much bulk to my bag, and if you leave 'em be, they can spread their spore love to grow more mushrooms. I also leave at least a third of a flush/patch of mushrooms behind to spread spores, for wild critters to munch on, or to leave a gift for a fellow forager. Also, tread lightly as to not trample little mushrooms. Be mindful of where you step so the fungi do not meet an unnecessary and early demise.

I like to trim and drop my mushrooms. What this means is, if I snatch up a mushroom, I'll cut off the dirty nub on the end, and if there are any bad spots I want to cut out, I let 'em drop onto the ground right where I found 'em to keep spreading the love.

I'm screamin' this again, Leave! No! Trace! ... Unless, of course, you're leaving it better than you found it. This includes picking up trash, not being a litterbug, or if you're rummaging through the leaves and such, pushing everything back the way you found it.

AIN'T NEVER NOT FORAGIN'

And finally, I ain't never not foragin', y'all, and you can do it, too. I advise you to turn any activity you possibly can into a foraging opportunity. For example, if you're on a long drive and you're a passenger, you can look to the trees and see if you can spot anything or even work on your tree ID skills. If you're at the park for a birthday party, keep that head on a swivel and see what you can see. If you're taking your dog for a walk, pay attention to the ground and look for some fungi. If you're bored and have some downtime, flip through your field guides and choose some new target mushrooms you wanna go after. Seize the moment to soak in your surroundings or beef up your skills.

CHAPTER 2

Know Your Parts (Mushroom Anatomy)

You may not have been expecting an anatomy lesson in a mushroom book, but here we are! Just like the human body, mushrooms have different parts. But don't ya worry, I'm gonna make it quick and painless.

Knowing the parts of a mushroom is super important. If you learn and know mushroom anatomy, it's gonna help narrow down your specimen during the identification process, and it's also gonna assist you in navigating this field guide. Comparing certain features of mushrooms will ensure you don't mistake your mushroom for a potentially dangerous lookalike. Some mushrooms may only have one or two features that separate the two, so if you're not privy to what those features are, it can sure make it difficult to differentiate.

LEFT | Did ya know mushrooms have their own body parts? It's true and apparently that can include a face, like this kinda angry grisette mushroom (*Amanita vaginata*). *Edible, but not recommended for beginners.

Here's an easy go-to reference diagram with the main mushroom parts I feel are need-to-know.

Now, let's take a deeper dive into these parts and look at some examples.

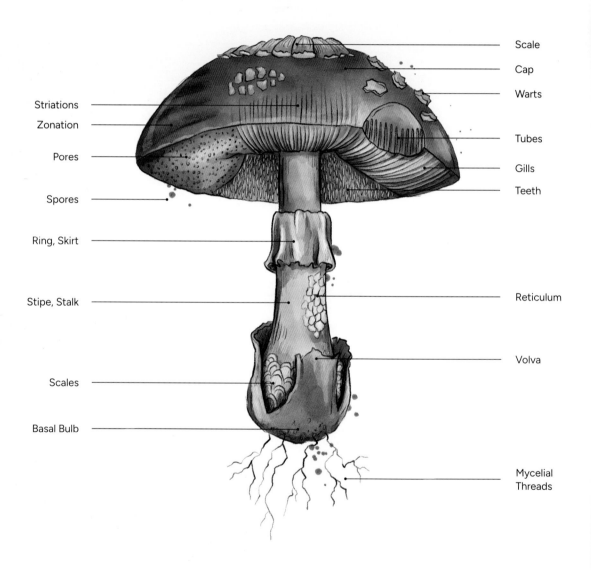

CAP FEATURES

The mushroom cap, also called the *pileus*, is the structure on top of the mushroom. Think of it like the mushroom's baseball hat. The cap houses the spore-producing surface of the mushroom. It can have gills, teeth, pores, etc. Caps can come in all different shapes, sizes, and textures (even slimy) and can carry a variety of features. Let's check out a few of 'em.

Warts

Some mushrooms have a protective sheath, called a *universal veil*, around them when they are youngins. That veil will break away as the mushroom grows up, and the process can leave little chunks on their cap. These speckly remnants are what we call *warts*. Warts are typically easily detached from the cap. I like to do what I call the flick test. If ya flick, does it stick? A quick flick should cause warts to detach and will send 'em flying off, whereas mushroom "scales" are more durable and not as flickable. When I think of warts on a mushroom, I think about the *Amanita* genus. One of the most widely recognized mushrooms anywhere is the fly agaric mushroom, or *Amanita muscaria*. It's that red Super Mario or Santa Claus mushroom with the white dots on top. You may have seen it featured on mushroom décor items everywhere. The fly agaric shows a perfect example of warts. The fly agaric mushroom doesn't grow in my neck of the woods, but its relative, the American yellow fly agaric, does. Just like its bubby, it too has warts.

These two different stages of maturation on the American yellow fly agaric (*Amanita muscaria* var. *guessowii*) exhibit how the universal veil leaves warts on the cap as it grows. ***This mushroom requires specific and extensive preparation to remove toxins that can cause unfavorable reactions. I do not recommend eating this 'un.**

Know Your Parts (Mushroom Anatomy)

Scales may also occur on the stem of a mushroom. These golden pholiota (*Pholiota aurivella*) show ya just that. You can observe their showy scales chillin' on both the cap and stem. ***Not edible***

Old man of the woods (*Strobilomyces strobilaceus*) is an edible shaggy feller that has blackish to gray scales across the cap.

Scales

Mushroom caps can also have a feature called *scales*. Scales become more pronounced in appearance as the mushroom gets larger, expands, and sort of kinda cracks. They serve to protect the mushie and can come in a load of different colors, sizes, and shapes. As I mentioned before when discussing warts, scales tend to be more durable and snugly adhered to the cap of the mushroom. They ain't gonna fly off when ya flick 'em. Scales will also usually be more three-dimensional in appearance (and sometimes spikey) than warts, which lay mostly flat against the cap's surface.

STRIATIONS

The margin of a mushroom cap can have what are called *striations*. Striations are radial, parallel, fine lines, or strips, that run around the circumference of the cap. They can sometimes resemble wrinkles and may appear to run right into the gills on the underside of the cap.

ZONATIONS

Some mushroom caps have zonations, which are differing areas, or "zones," of varying colors and/or textures. Think of them as alternating beauty stripes, like on a zebra.

The woolly milkcap (*Lactarius torminosus*) wears very distinct and alternating zonations of color and texture. *Not edible

This grisette mushroom (*Amanita vaginata*) is bowing down and just beggin' for ya to peep its purdy striations. **Edible, but I do not recommend consuming mushrooms in the *Amanita* genus for beginners due to potentially poisonous (and even fatal) lookalikes.**

Turkey tail (*Trametes versicolor*) can exhibit a variety of color zonations in shades of blues, browns, blacks, reds, creams, and whites.

SPORES AND SPORE-PRODUCING SURFACES

In the diagram on page 24, you saw some little sprinkles coming down from the mushroom labeled "spores." Think of spores like microscopic mushroom seeds. Singular spores are not visible to the naked eye. Spores are teeny, tiny reproductive cells that float around to spread the mushroom's genetic material to promote further growth and reproduction. So, the mushroom is literally spreadin' its seed! These spores go on a fungi journey. They attach to surfaces in an attempt to colonize and grow new mushrooms in new locations.

The spore-producing surface is the part of the mushroom that releases those lil' mushroom seeds into the world, and you'll typically find that surface located on the underside of the mushroom cap. Mushrooms can have different spore-producing surfaces that look quite different from one another, including gills, pores, and teeth. There are other mushrooms that don't quite fit into the gill, pore, teeth categories, including morels and other "misfits."

Gills

When you think of a random mushroom in your head, that mushroom probably has gills on it. This makes a whole lotta sense since gills are one the most commonly occurring spore-producing surfaces in the mushroom world. Gills are thin, flimsy structures that hang vertically under the mushroom cap. These gills produce those spores we talked about before.

In a utopian world, it would be as easy as, "Does the mushroom have gills?" You'd answer yes or no and move on in the ID process. It's a little more in depth than that. Not all gills are the same, and the gills themselves can have their own characteristics. Let's have a look-see at some different gill types you might stumble across.

Decurrent gills run down the length of the stems on this cute lil' oyster mushroom cluster.

When you're looking at gills, they can be what is called *decurrent*, *free*, or *attached*, which references how the gills are attached to the stem. Decurrent gills mean the gills continue to run down the stem. Oyster mushrooms are an excellent example of a mushroom with decurrent gills. In the photo above, you can see and follow the gills as they continue to run down onto the stem of the oyster mushroom.

Free gills do *not* attach to the stem of the mushroom at all. There will be a noticeable space

This sure was a first. This little yellowfoot fungus with forky gills sprouted straight from a hemlock cone, proving that sometimes mushrooms have no rules.

This deer mushroom (*Pluteus cervinus*) showcases its free, or detached, gills while I showcase my best mushroom huntin' manicure. You can observe the distinct gap between the gills and stem and the distinct chips in my nail polish.

between where the gills end and the stem begins. In a younger mushroom specimen, it may be a tad bit harder to see that space versus an older mushroom specimen, but it will still be observable with a closer look and squint of the eye. Many of the mushrooms you can readily purchase at the grocery have free gills, like button mushrooms or portobellos.

There are also some mushrooms with attached gills, but they may not be necessarily decurrent. This means that the gills *will* touch the stem, but not keep truckin' on down any length of the stem.

There are *adnexed* gills, which mean they are narrowly or barely attached, with only a small part of the gill meeting the stem. There are also *adnate* gills that run into the stem and are broadly or squarely attached.

Gills can also be what's called *forked*. This means the gills of the mushroom branch off or have a pronglike appearance to them. On this page is a picture of a yellowfoot mushroom showin' off its forky gills. Notice how they branch off, especially as they approach the margin of the cap.

TRUE VS. FALSE GILLS

Mushrooms can have true or false gills, but don't let this intimidate you. It's quite simple to tell the difference after you see a few examples.

True gills are the quintessential light, feathery gills. They are papery and thin in texture. When you run your finger across them, they will be soft and move around freely. I like to do a fingernail test. If you take your fingernail and rake it across the gills and they break easily, they are likely to be true gills.

False gills are more like folds or raised ridges. Think of them more as wrinkles or veins. They are blunt and more bumpy-ish. When doing the fingernail test, they will not break or move freely like true gills do.

When I think of a mushroom with false gills, I think of the tasty chanterelle. In the summer when those golden nuggets are popping up everywhere, you may also come across a potential lookalike called the jack-o'-lantern mushroom (*Omphalotus illudens*). You do NOT want to eat the jack-o'-lantern mushroom. It is toxic, and though it may not kill you, you'll wish you were dead after you puke your guts out and can poop through a door screen after eating it. There are a few ways to tell the difference between the edible chanterelle and the inedible jack-o'-lantern mushroom, which I will cover more in depth in the chanterelle mushroom profile on page 72, but I wanted to bring to the forefront here the importance and difference in the gills of these two mushrooms. The jack-o'-lantern has *true* gills. The chanterelle has *false* gills. This is a surefire way to confirm your identification. The two photos below are examples of the true gills of a jack-o'-lantern mushroom and the false gills of the chanterelle side by side. Notice the delicate gills of the jack-o'-lantern and the more sturdy, raised ridges of the chanterelle. You can find even more information on how to ID a chanterelle from jack-o'-lanterns later on in chapter 6!

Jack-o'-lantern mushrooms have light and delicate "true" gills. They will have a feathery feel to them and can be easily moved around or broken. *Not Edible

Chanterelle mushrooms have sturdier raised ridges known as "false" gills that are more wrinklelike. False gills are not easily moved around or broken.

BRUISING

When you are lookin' at this field guide or any other mushroom literature and it mentions that the mushroom's gills exhibit bruising, what that means is that when the mushroom is cut, nicked, or damaged, the gills can change colors. This is an important feature when you are identifying milk cap mushrooms. When you cut the gills of a milk cap mushroom, you may see the gills are left bruised or have a color change where you cut. One of my favorite milk cap mushrooms to forage is the tawny milk cap (*Lactifluus volemus*, see page 90). Their gills will usually be a pale-yellow color, and once you cut those gills with your mushroom knife, they excrete a white liquid (milk). You'll see the gills will then bruise and turn a brownish color where the mushroom was cut. That's the bruising! There's another milk cap mushroom called the saffron milk cap (*Lactarius deliciosus*) that bruises green after being cut.

IS IT MILKY?

I mentioned milk caps in the previous section on bruising, but I wanted to further cover that milky topic. Milk caps or *Lactarius* mushrooms are mushrooms that ooze a milky substance. You can check whether or not your mushroom is a milk cap by simply givin' it a fresh slice. If it drips milk, it's a milk cap! The "milk" it secretes is called *latex* in the mushroom world, but it's not true latex. It's just been coined that, so you can call it milk or latex—whatever floats your boat. Not all milk caps are created equal. Dependent on the species and age of the mushroom, it might be a gusher and produce a whole lotta milk, or it can produce a very small amount. The milk of *Lactarius* mushrooms comes in a whole rainbow of colors, some secreting red to yellow milk to even blue or purple milk. Some milk can be clear and watery or cloudier and thicker. The milk can also carry a distinctive scent, which is another good way

The corrugated milk cap (*Lacatarius corrugis*) slowly bruises brown after it is cut and releases its milky goodness.

Indigo milk caps (*Lactarius indigo*) excrete a blue/purple–colored mushroom milk.

to identify what ya got. The lower image on this page is an example of a *Lactarius* mushroom, the indigo milk cap (*Lactarius indigo*), lettin' go of some of its milk. Notice the blue "milk" on the blade of my mushroom knife.

GILL SPACING

Different mushrooms can have differing gill spacing, which just means how close together or far apart the gills are. They can have *crowded* gills, meaning the gills are packed tightly and very close together; *close* gills, meaning they are close together; *subdistant* gills, meaning they are spaced a little further apart than close gills; and finally, *distant* gills, meaning they are widely spaced apart. Below is a diagram to give you an idea of these different gills spacings.

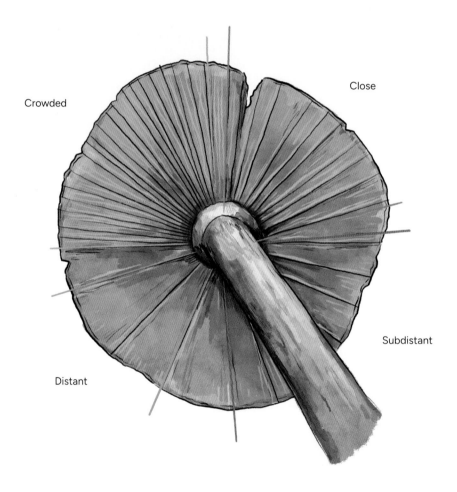

Pores

Pores are another type of spore-producing surface a mushroom can have. Pores are little holes on the underside of the cap.

Boletes are a common mushroom with pores. The underside of a bolete is much like a sponge, with itty, bitty, teeny, tiny holes, visible with your nekkid eye. Those holes are the pores! Mushrooms with pores also have what are called *tubes*. When you cut a porey mushroom cap in half (cross section), you expose small, cylindrical tubes that run down inside the cap to the pores. Mushrooms are so weird and cool. The tubes allow the spores to travel down and bust out into the world through the pores to spread their mushroom legacy.

There are also mushrooms referred to as *polypore* mushrooms. Like boletes, they have small pores on the underside of their caps (and tubes inside of them). Polypore mushrooms differ from boletes in that polypores are typically tough, leathery mushrooms that grow in overlapping patterns. They also often like to grow from wood. Some examples of polypore mushrooms include chicken of the woods, turkey tail, pheasant back, and beefsteak polypore.

Teeth

Another spore-producing surface a mushroom can have is teeth. Yes, I said teeth. Sounds crazy, huh? Toothy mushrooms are some of my favorites to find because they are so fun to look at it. Teeth are spikelike, dangling structures that hang down from the underside of the mushroom's cap (if present). Teeth occur less frequently than gills and pores, but that's not to say there aren't plenty of yummy, toothy mushrooms to be had. Some examples of edible mushrooms with teeth would include lion's mane, bear's head tooth, coral tooth, and my favorite mushroom out there, the hedgehog mushroom!

Boletes have a spore-producing surface made of tiny holes, called *pores*, as seen here on this ornate-stalked bolete (*Retiboletus ornatipes*).

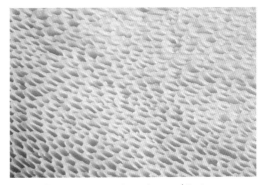

Pores of a pheasant back mushroom (*Cerioporus squamosus*).

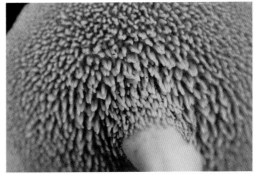

Hedgehog mushrooms (*Hydnum* sp.) have teeth on their undercarriage.

Morels and Other Misfits

As you've probably figured out by now, mushrooms love to bend and break all the rules. So, not every mushroom is gonna have a spore-producing surface that fits neatly into the gills, pores, or teeth category.

There's a group of mushrooms called *Ascomycota* mushrooms, sometimes called *sac fungi*. The morel mushroom is in this family and has a spore-producing surface located on its honeycomb-lookin' cap. The morel does not have typical pores, gills, or teeth, but a spore-producing surface all its own. Morels can be so extra.

Ascomycotas also include **cup fungi**. Cup fungi grow in the shape of a cup, bowl, saucer, or goblet, and the spores are formed and dispersed from the inner surface of the fruiting bodies. The spores are spread through raindrops hitting the inside and splashing or through the wind blowing them around. Something pretty cool about finding certain cup fungi is that you can blow into the cup/bowl of the fruiting body and then in just a second, you'll see the spores fly out in a little cloud. Some cup fungi audibly hiss or pop when they release these spores. Again, mushrooms are so weird, and they rule. This is why we love them.

Ascomycota mushrooms also include **truffles**. These are the highly sought after roundish, bulbous guys that grow under the ground. You may have seen them being dug up by truffle-sniffing pigs or dogs. This is how truffles like to spread their spores. In their more mature state, they release a strong scent to seduce certain animals to dig 'em up and eat 'em, which spreads the spores. Some types of truffles are so dang expensive due to their scarcity. They are hard to find, hard to cultivate, and don't have a very long shelf-life. The average cost of black truffle is approximately $1,000 to $2,000 per pound! White truffles are even more expensive, with a pound costing as much as over $7,000! That doesn't really jive with my hillbilly pocket change, but I have been lucky enough to try a little bit of truffle in my day, and I will say, it's pretty darn tasty. Most black and white truffles are unearthed in Europe. There are such things as Appalachian truffles (*Tuber canaliculatum*) that are being talked about more as of late here in the United States, and I hope one day to get my hands on one!

The common brown cup (*Peziza phyllogena*) is another cup fungus. If you blow onto them, you can see and maybe even hear them release spores!

Devil's urn (*Urnula criterium*) is a primo example of cup fungi. You can really see how that name came to be with these fellers.

Behold a stunnin' and pricey black summer truffle (*Tuber aestivum*).

These common puffballs (*Lycoperdon perlatum*) can be found all over the dang place in the late summer and fall.

There is another family of mushrooms called Basidiomycota that includes puffballs, stinkhorns, jelly fungus, coral, and club fungi. I bet you'll stumble across these in your travels, and you'll see they don't have gills, pores, or teeth.

You might already be familiar with the **puffball mushroom**. There are a few different types of puffballs, but these are the mushrooms that grow in a ball shape, and once they're a little older, give 'em a lil' squeeze and they will "puff" out their spores from a hole that opens on its surface.

Stinkhorns are a sight to be seen. There are a few types that can vary in shape from cylindrical to even a cagelike appearance. You might believe you've come across an alien lifeform sprouting from the ground. The common stinkhorn is the one you'll most likely encounter. And boy, howdy! They will probably shock ya at first. They are, um, rather phallic in appearance and like to surprise ya when they pop up in your flower or garden beds. They get their name because they smell plumb rancid. They can kinda resemble a half-free morel but give it a sniff if you're unsure. If you gag, it's a stinkhorn. They are a neat mushie because the stinky slime on their cap attracts flies. The flies crawl all over it and get spores on their lil' feet. This is how the stinkhorn spreads its spores.

These raunchy stinkhorns are in differin' states of maturation. The one on the left has had its goop eaten off, and the one on the right is ripe for the pickin' for the bugs.

Scan here for a lil' bonus truffle lasagna recipe!

Know Your Parts (Mushroom Anatomy)

Snow fungus (*Tremella fuciformis*) is an example of some glistenin', wiggly and jiggly jelly fungus belonging to the Basidiomycota family. **Edible and medicinal*

This yellow-tipped coral fungus (*Ramaria formosa*) exhibits perfectly how this family of mushroom got its name. **Not edible*

These dead moll's fingers (*Xylaria longipes*) are a type of club fungi you can find fruiting from wood. **Not edible*

Jelly fungi are appropriately named and really easy to identify as jellies when ya find 'em. They are rubbery and have the consistency of their namesake—jelly. They can be gelatinous, brainy, wet, sticky, or maybe even dried up and hardened from the elements. If you give them a dunk in water, they reconstitute and return to their original glorious form. You will find jelly fungi growing directly from decaying wood. A few prized edible jellies you can find in the eastern United States includes wood ear (*Auricularia auricula-judae*), amber jelly roll (*Excidia recisa*), witches' butter (*Tremella mesenterica*), and snow fungus (*Tremella fuciformis*).

Coral fungi are another appropriately named fungus because they look like, you guessed it—coral! Think of those pretty, deep sea structures you'd see down on the ocean floor. These funky guys are made up of clusters with branches or "fangers," as I like to call them. Coral fungi come in a rainbow of colors and have tons of fun shapes. The individual fangers of the cluster can be clublike or even forked with tiny prongs on top. They spread their spores through their fangertips, shootin' that mushroom magic all over the place. They are mostly found growing from the ground but certain species can also fruit from dead/dying wood.

Club fungi are a type of fungus that comes creepin' up from the earth or poppin' out from wood. They are club-shaped, but ain't as branchy as their sisters, the coral fungi. They can look quite ominous and eerie, especially one known as dead man's fingers (*Xylaria polymorpha*) that looks like a dead feller is diggin' his way out of his grave below. Club fungi aren't all straight from a horror film, though. They can be brightly colored and cheery, too!

STIPE/STALK/STEM

If a mushroom has a stipe/stalk/stem present, it is the easily observable, often elongated shaftlike structure of the mushroom that supports or holds up the cap. Think of an umbrella when you think of a mushroom here. The cap of the mushroom is the top wide part of the umbrella that spreads out and keeps the rain off you, and the stem is the cylindrical handle that holds the top of the umbrella up. The stem itself can have a few different features we will cover in the sections to follow.

Universal/Partial Veil

Some mushrooms have what is called a *veil*. This is a protective layer or blanket that safeguards the mushrooms when it's just a lil' baby fungus. There can be a *universal veil*, which will cover the entire mushroom-fruitin' body, or there can be a *partial veil*, which will cover a portion of the mushroom, usually the spore-producing gill surface. Think of a veil on a bride. A bride wears a veil to protect the goods and hide her from evil spirits that may try to do her harm or obstruct her happiness. This is similar to the veil on mushrooms, keepin' any evildoers away from interfering with the mushroom's glow up. As the mushroom continues to mature, the veil eventually opens or breaks away, exposing the beauty of the mushroom.

Skirt/Ring/Annulus

Mushroom stems can wear what I like to call a *skirt* or a *ring*. You may also hear it referred to as an *annulus*. The skirt is actually the remnants of a partial veil that have held onto the stem as the mushroom grew up. Let's go back to the umbrella example. The partial veil that once covered up the spore-producing surface of the mushroom breaks away as the mushroom grows up and out. So, much like when you open an umbrella and it goes from its closed form to its open form, this is how that mushroom has grown. Pieces of the veil cling to the stem, creating the skirt. You'll usually find it where the cap once met the stem, on the upper portion of the stem. Identifying whether the mushroom does or does not have a skirt is very important in mushroom identification.

Many mushrooms in the *Amanita* family have a veil in their younger growth states. Here, you can see the protective layer it provides to the gills under the cap. You can also see how it has started to break away and leave a "skirt" on the stem. I do not recommend eating mushrooms in the *Amanita* family for beginners.

Stickin' with the umbrella theme, here's a turtle I encountered seekin' some solace under the shade of a mushroom umbrella!

The edible meadow mushroom (*Agaricus Campestris*) has a ring on the stem that is the remnants of the partial veil that once covered the gills when it was a younger fun-guy.

The stem of the shaggy stalked bolete (*Aureoboletus betula*) shows a netlike pattern of nooks and crannies known as reticulation. *****Edible**

Reticulation

Mushroom stems may also be "reticulated." *Reticulum* refers to a weblike or netlike pattern adorning the stem. It appears to have a wrinkly, raised, and pitted surface. Reticulated stems are commonly seen in bolete mushrooms.

Volva

Some mushrooms have what is called a *volva* at the base of their stem. The volva is the leftovers of a universal veil that once completely encompassed the mushroom in its earlier stages. The volva looks like an egglike sack from which the mushroom hatched. The volva serves as a comfy, safe place for the baby mushroom. You might be frolickin' in the forest and find what looks like a random alien egg just sitting on the ground. This could be a universal veil of a mushroom, soon to be a volva, that just hasn't unleased the fruiting body yet. When the mushroom hatches or breaks through the "egg," it can leave the remnants of the volva behind on the ground, which can be a major factor in identifying that mushroom. Depending on the age of the mushroom, the volva may be very apparent, or it may not. If you suspect the mushroom should have a volva, ya may wanna dig down a little deeper and take a closer look to see if a one is present.

This Eastern Caesar's Amanita (*Amanita jacksonii*) hatches out of a volva you can see here at the base of the mushroom. *****It's edible, but not a beginner mushroom!**

Basal Bulb

Some mushrooms have a basal bulb on 'em—or what I like to call a big ol' booty. The basal bulb is found on the lower part of the stem as a round, bulbous, or swollen mass. The bulb will be wider in diameter than the stem located above it. The basal bulb is connected to the mycelium. Sometimes the big ol' booty has scales on it, and sometimes it doesn't. Check for this when identifying, and luckily, you already know exactly what scales are and what they look like! In order to see if the mushroom has a basal bulb, you might have to dig a bit and excavate the fruiting body wholly and carefully from the ground. You can then take in its full voluptuosity and use this to confirm or deny your ID.

This *Amanita* mushroom let's ya know it's got a big ol' basal bulb right off the rip. There's no need to excavate here to determine that. *****Eating *Amanitas* is not recommended for beginners**

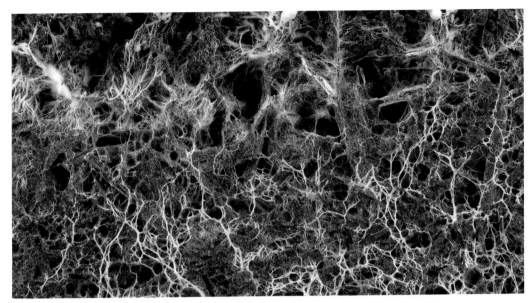

Mushroom mycelium looks a heck of a lot like spiderwebs as it weaves through its substrate.

MYCELIUM

Finally, let's talk about the brain of the mushroom world—the mycelium! This stuff is amazingly interesting and so important. It is the largest organism in the world, honey! Mycelium can kinda be thought of as the root system of a plant, but way cooler because it's mushrooms. Mycelium is located everywhere below your feet, underground, as a massive, branching system made up of teeny fungal strands called *hyphae*. When you are lucky enough to see mycelium (most of the time it is microscopic), it may appear like a white, webby material. The mycelium itself is not a mushroom, but it is where the aboveground fruiting bodies of fungi we call mushrooms are generated from. I always compare this to an apple tree. The vegetative structure of the apple tree would be like the mycelium, and the apples are the mushroom fruiting bodies.

Mycelium gives nutrients and life to mushrooms. The mycelium can branch out and attach itself to different root systems of plants to gain and swap nutrients between the fungal hyphae and the plant or tree. This is why we find certain mushrooms around certain types of trees. They form a relationship with specific trees through their mycelium in order to thrive.

When harvesting mushrooms, it is important to not damage the mycelium. There's no need to take a backhoe out or plow up the ground to get a mushroom. A simple cut, pluck, or shallow dig works just fine to harvest while keeping the mycelium intact and workin'.

CHAPTER 3

Common Mushroom Myths

Something I have encountered many a time since I started doin' this whole mushroom thing are naysayers and wet blankets out to squash our fun. Maybe they are misinformed, but I always say, do your research before you spread any negativity about my mushroom babies! There are lots of myths floating around out there that are just completely ridiculous and may deter people from foragin' for fungus. I wanted to address a few of the common mushroom myths I've heard the most often. I'm gonna debunk them fer ya so your head ain't filled with any of the nonsense that perpetuates unnecessary fungi fear.

DON'T TOUCH 'EM!

Mushrooms can sometimes get a bad rap. They unfortunately are oftentimes related to death, poisoning, and toxicity. There is a fear called *mycophobia*, which is the fear of mushrooms, and it is very real for some folks. When the show *The Last of Us* came out, where mushrooms turned people into zombies, it didn't help the pro-mushroom cause any, that's for sure. In reality, mushrooms are really nice fellers. Many folks were raised up to think you should never touch a wild mushroom or you'll poison yourself and die. I know this still rings true today because I have received tons of ID requests in my DMs and the pictures people send are sometimes of them holding a harmless little mushroom in a pair of heavy-duty welding gloves or a full suit of armor. I can maybe understand this if people grew up being scared to death of mushrooms, but the lengths people go to to not touch an innocent lil' fungi will always tickle me. The truth is, you shouldn't be afraid to touch mushrooms. They won't bite and they won't poison you from simple contact with your hands. Toxic or poisonous mushrooms must be *ingested* to cause harm. I have no problem walkin' up to a destroying angel mushroom (*Amanita bisporigera*), one of the most toxic mushrooms in Appalachia, and givin' it a lil' pet or a gentle cradle in my arms. Some people recommend washing your hands afterwards, and if that makes you feel better, have at it, buddy! I will suggest that if you find a mushroom you may suspect to be toxic and want to bring it home for further inspection, do not put it into your edible mushroom sack. Keep these questionable mushies in their own separate place to make sure no broken bits get mixed in with the edible and confirmed-safe mushrooms you want to eat later.

To further confirm how not scary even toxic mushrooms can be, I wanna talk about something

Destroying angels (*Amanita bisporigera*) are a common toxic mushroom that you can touch and won't die. ***Just don't eat it, and you'll be fine.***

*LEFT | Although mushrooms can have teeth like this lil' lion's mane (*Hericium erinaceus*, edible), don't worry, they won't bite ya.

called the *chew and spit test*. The chew and spit test is something some folks do in the field to confirm/deny their mushroom ID. Some mushrooms—upon a raw nibble and chew—yield a distinct flavor in your mouth, such as bitter, spicy, or sweet. Once you chew it a bit, you then spit out the specimen and *do not* swallow it. This can be done even with poisonous mushrooms. Do *I* do this? Not usually, but can it be done, yes! I don't recommend the chew and spit test for beginners at all. Me telling you about it is just to show you that mushrooms aren't as scary as your parents may have led you to believe.

With all this being said, the only mushroom to my knowledge that you *maybe* shouldn't pick up would be fire coral fungus (*Trichoderma cornu-damae*), which grows in Asia. It can *possibly* cause skin inflammation. That's still up for debate. So, if you're plannin' on skippin' through the continent of Asia for a forage and you see a red coral fungus springin' up from the ground, maybe don't touch that one.

Fire coral fungus (*Trichoderma cornu-damae*)

GILLS KILL!

If I had a nickel for every time someone commented on one of my social media posts to say "Gills kill!" I'd have a whole lot of nickels. I don't really have much to say about this other than it's just not true. A mushroom with gills does not equate to a mushroom that will hurt you. Some of the most delicious, safe, and edible mushrooms have gills, such as oyster mushrooms, milk caps, and enoki, to name a few. So, don't let anyone trick you into being wary of eating edible, gilled mushrooms, and tell them I said they ain't right in sayin' so!

YOU'RE HURTING THE MUSHROOM POPULATION!

Here's another comment I get a lot. When I find a big cluster of mushrooms and take a haul home, I sometimes get accused of messin' up that patch of mushrooms. Truth is, I ain't causing any problem in making my harvest. As discussed earlier, mycelium is the good, good stuff that lives underground and produces the fruiting body that is the mushroom I pick. Again, think of the apple tree. The mycelium is the tree, and the mushroom is the apple. Picking the apple does not kill the tree, and the tree continues to produce apples each year. Picking the mushroom doesn't kill the mycelium, so it's still gonna spit out mushrooms in the future. Just remember to not disturb the mycelium underground and you're good. I always leave some mushrooms to spread spores, and I never take more than I intend to use. Don't be greedy and only take as many as you need so as not to waste. It's also nice to leave some for other folks and forest creatures to munch on.

Tawny milk caps (*Lactifluus volemus*) are one of my favorite summer milk caps to munch on, and they are gilled mushrooms! Gills, in fact, do not always kill.

Talk about a good day with this haul of chanterelle mushrooms! I have a honey hole I visit every year and clean up on these babies in the summer. This goes to show you *can* responsibly bring a haul home and return to that spot each year and get more. No harm done. Just be respectful to your honey holes and to Momma Nature in general, and she will treat you right!

I like to use my mushroom knife to cut shelf mushrooms, such as this gorgeous chicken of the woods (*Laetiporus cinnicinatus*). But, if I felt like it, I could just pluck it straight from the wood. It doesn't matter! Choosing cuttin' or pluckin' does not positively or negatively impact future mushroom growth.

CUT! DON'T PLUCK!

Ya ever heard of "pick shaming"? If not, this is when someone criticizes others for how they harvest their mushrooms. Some people solely cut the mushroom with a knife. Some people grab and pluck the mushroom from the ground or wood. You'll have some folks who swear that cutting the mushroom is healthier and helps the mushroom to come back better than plucking it. The truth is that studies show squat. Neither way has been shown to be superior to the other. Contrary to the belief of some, the mushroom does not regenerate after decapitation or cutting. A mushroom will not spring up from a stump you leave behind. I honestly use both methods depending on how I'm feeling. You may wonder why I even carry a knife then if I can just pluck. The reason I bring my mushroom knife along is that I love being able to use the mushroom brush on my knife to clean 'em up. I also like to have my mushroom knife to more easily cut mushrooms growing off wood. And sometimes, I do like to use it to cut them from the ground. Like I said, if the spirit moves me one way, I'll cut, or if it moves me the other way, I'll pluck. It really doesn't matter. You cut or pluck your little heart out, honey! Whatever feels right to a ya. And don't listen to the haters.

Go Forth and Forage

When it comes to identifying mushrooms, you have to ID with confidence as there is no one way to tell if a mushroom is toxic or not. Using your mushroom guides in the field is a great way to determine what ya got.

HOW CAN YOU TELL WHAT'S POISONOUS AND WHAT'S EDIBLE?

Finally, how do we tell if a mushroom is poisonous or edible? Surely there must be some way to do this, right? Well, no, there ain't. It would be nice if there was some general rule of thumb to follow, but there just isn't, pal. I've said it once and I'll say it again: Mushrooms play by their own rules sometimes and keep us on our toes, so it makes sense they don't make it easy on us. ***No single thing can determine whether a mushroom is edible or poisonous.*** Some people may see a brightly colored mushroom and think, "Oh goodness! It's a wild color, so it must be poisonous!" Nope. That's not how it works. In order to know whether a mushroom is edible or inedible, you just gotta know your stuff. You have to confirm the identity of that particular mushroom with 110 percent confidence and know what it is in order to know whether you can swaller it safely or not.

CHAPTER 4

Words of Wisdom

I'm not ordained or anything, but I do know me a thing or two about mushroom huntin'. So let me preach at ya right-quick from the pulpit. I wanna share some final and overall words of "Whitney Wisdom" with ya before we take a deeper plunge into the mushroom profiles. Hopefully, these shroomy sermons of mine will get ya good and comfy so you're better ready to go forth and forage!

START SMALL AND NOT RAW

Some wild mushrooms are better tolerated by some folks more than others. Even the most perfectly safe, edible mushroom can be eaten and enjoyed by one person but cause another person to get the bubble guts. This is why I always recommend when trying a new-to-you edible mushroom, to start *small*. By this I mean try only an itty-bitty portion of the cooked wild mushroom to see how your body is going to react. If you feel icky and a strong urge to hit the toilet, maybe it's a mushroom that doesn't agree with you. You'll sure be thankin' your lucky stars you only had a few bites instead of a supper plate full if it doesn't go the way you had hoped. I'm gonna share a secret with ya. I am unfortunately plagued by an inability to eat a certain chicken of the woods mushroom called *Laetiporus sulpherus*. That species makes ya girl real sick like and apparently can be problematic for some other folks, too. I had foraged some on vacation years back and cooked it up with scrambled eggs for pre-hike breakfast. I was not able to do any hiking or much of anything the rest of the day other than post myself up in the bathroom. I'll admit, it made me super sad, and I felt like a phony or somethin' for not being able to eat one of the most sought-after wild mushrooms. I am stubborn and thought maybe it was just a fluke, so I tried another harvest of *Laetiporus sulphureus* a few weeks later. Sick again. Don't be like me. You're not a phony if you can't eat a certain mushroom. Listen to your body and be smart! I have learned my lesson since then. I found out after the great diarrhea blowout incident of 2020 that I can in fact eat the other type of chicken of the woods, *Laetiporus cincinnatus*. Apparently, this species is better tolerated across the board by people, and it happens to be the preferred variety of chicken of the woods anyway. That helps my psyche and my pride. Ya win some and ya lose some.

So, we've covered why you should always start small with foraged mushrooms, but I also suggested **not to eat a new-to-you mushroom raw**. As a general rule of thumb, I always preach to give your mushrooms at least 10 minutes on some heat to cook thoroughly. Mushrooms have what is called *chitin*, a part of the fungal cell wall that is hard for us humans to digest. But a good ol' heat preparation gets that gone for ya. Cooking your mushrooms also breaks down some toxins or compounds that can cause potential tummy rumbles. Morels are a superb example of this. Morels are a perfectly safe, tasty, edible mushroom—when cooked. If you eat raw morels (but just don't), buckle up for some GI woes and possible not-so-fun neurological symptoms. Of course, there are *some* exceptions to the rule, and some mushrooms *can* be consumed raw. However, it's best to err on the side of caution, especially as a beginner. Don't risk it. Cook them babies up!

And of course, I'll keep screamin' it until I'm blue in the face: Don't ever eat a wild mushroom unless you're so, so, so sure that it's a safe, edible mushroom.

LEFT | Turns out the sun does shine in the crack of an oyster mushroom. (*Pleurotus ostreatus*, edible).

DO NOT GET DISCOURAGED

The holler wasn't built in a day and neither is masterin' the art of mushroom foraging. I still ain't "mastered" it and probably never will. I don't think anyone ever does. There's always something to be learned and new things to be seen. I'll admit, it can be a lot of information to take in at first and ya might feel like peterin' out at times. I'm here to promise you just one thing: You *will* get there if you want to! We all gotta start somewhere, and the bottom is the best place to work your way up from. It does take plenty of work, research, practice, help from others, and gettin' your butt out there to get yourself to a place where you can confidently identify and eat wild mushrooms without fear. I've been working at this thing for a while now and I am living proof—if you want to put in the effort, *you can do it*. When you love something and you're passionate about it, the work you put in and the research you do does not seem like a grind, and it is actually downright enjoyable. Every day I hit the hills, I am plum gassed up to apply my learned skills and to expand my repertoire with each foray. Just keep at it and don't expect perfection overnight because that is terribly unrealistic and unattainable. Enjoy the ride because it's so fun workin' your way to the destination.

BE GOOD TO YOUR MOMMA (NATURE)

We touched a lil' bit on this already in the responsible and ethical foraging section earlier, but it is much deservin' of a round two clapback because it is so daggone important! In order for us to continue to enjoy the gorgeous outdoors and all of its offerings, we have got to show Momma Nature the care and respect she is more than oh-so-worthy of. This can be done in a multitude of ways, and I hope you can adopt some of these practices to beautify your stompin' grounds if ya ain't already.

- Don't litter the land. Being a litterbug is the most uncool thing you can be as a forager in my opinion. You probably don't like people comin' into your house and makin' a mess and neither does Momma Nature. I encourage you to pick up trash or anything that doesn't belong in the woods whenever you can. Minimize your impact. If you packed it in, pack it out! Leave no trace or leave it better than you found it.
- Don't rape the land. Take only what you'll use. Don't be greedy. Leave some for other people and animals. Leave some to promote future growth.
- Don't destroy the land. Don't trample things. Don't go cuttin' down trees, breakin' branches, strippin' bark, or uprootin' things for no valid, legitimate reason.
- Be privy to any and all protected/endangered/rare species of flora out there and harvest regardfully.
- Respect the wildlife.
- And always remember to show your appreciation for Momma Nature. Tell her thank you.

YA AIN'T GOTTA BE FANCY

Foraging does not require bells, nor does it require whistles, friend. Ya don't need no name-brand, expensive duds or any high-falutin' gear to get the job done. I ain't got any of that, and I think I'm doin' mighty fine so far. You don't *have* to have specific gear if you don't want it. You can go out with your bare hands and a tater sack to forage mushrooms. I've been out and about before and stumbled upon an unexpected patch of goodies without a mushroom bag on me. I've used my shirt to fashion a hillbilly mushroom sack to pack chanterelles in. I've filled the hat I was wearing on my head with enoki mushrooms before. I've tucked oyster mushrooms straight into my purse. If you don't have a mushroom-specific knife, that's okay! It ain't gonna make or break your efforts. Just grab those mushrooms. I will say, if mushroom foraging is something you plan to do regularly, I do recommend at least looking into getting a bag for your planned forays. As mentioned in an earlier chapter, there are very affordable mesh bag and basket options available online to make your life easier. And you'll be doin' your part to spread the spores!

Part of the merriment of mushroom hunting for me is the sweet added bonus of it having the "free and cheap" aspect to it. It costs zero dollars and zero cents to go outside, and you don't even have to travel far to find some fungi. The stuff is seriously everywhere! I often find wild food just a few steps from my porch. Even if you live in a bigger city with less wilderness surrounding you, there's mushrooms to be had if you just take a closer look. You won't spend a penny when you're pickin' a wild mushroom. Sure, maybe you'll have to spend some money on gas if you decide to drive to a trail, but compared to other activities or hobbies, it's purdy darn cheap.

DON'T PUT THAT IN YOUR MOUTH!

I done told ya, but I'm gonna beat a dead horse here: *Never, ever eat a mushroom you are not completely, totally, utterly, entirely, fully, and wholly sure is an edible and safe mushroom to eat.* If you're skeptical at all, don't put that in your mouth (which is also a good philosophy to apply in all areas of your life). Seek only trusted people and resources to confirm your identifications when you need a hand.

HAVE A BIG OL' TIME!

Foragin' is a blast, a rush, a thrill, a joy, and it doesn't need to be anything else. Don't stress yourself out or sweat the small stuff. It ain't worth it. If you find yourself not havin' fun while you're learning and foraging, be kind to yourself and step away for a hot second. Give yourself the healthy, deserving space you need to learn at your own pace. There are no deadlines or time crunches! We don't want this to turn into something that isn't absolutely gratifying. Oh, and don't ya dare compare yourself or your skills to others because you are doing *your* best, sweetie, and that's what counts. Keep on truckin'!

Another thing to remember is that you're not always gonna find a big ol' haul of mushrooms every time you go out. Don't let this knock ya down. Finds are not guaranteed. If they were, it wouldn't be as much fun when you do find somethin' good, now would it? On those inevitable days that you go home empty-handed, be mindful and recognize that you got outside, soaked up some nature, exercised both your body and mind, and hopefully had a big ol' time doin' it!

Steps down from pulpit

NOW, LET'S DO IT TO IT, FORAGIN' FRIEND!

Even the smallest finds, like this tiny lion's mane (*Hericium erinaceus*), will send me into a screamin' fit of excitement. All fungus finds, no matter how big or small, are a win in my book!

PART 2 — EDIBLE APPALACHIAN MUSHROOMS BY THE SEASON

Bless your heart, you've made it! Let's go over a quick breakdown of each mushroom profile you're about to feast your eyes upon.

- **Looks** will be a description of that mushroom's appearance: its size, shape, color, patterns, textures, and the like.
- **Stomping Grounds** will discuss where that mushroom likes to grow: what trees, terrain, conditions that mushroom tends to be with.
- **Growth Pattern** will go over *how* that mushroom grows: singularly (by itself), scattered (here and there), gregariously (in groups, but individually, not coming from the same base) in clusters, from the soil, on wood, and so on.
- **Spore Print Color** lets you know what color the spore-producing surface should give you when you do a print.
- **Flavor Notes** will describe the taste of the mushroom.
- **Potential Lookalikes** will go over other mushrooms that may appear to be similar to the mushroom being discussed, along with information of characteristics and features to assist you in telling the difference.
- **Right-Quick Rundown** will be a "right-quick" inventory of that particular mushroom's key identifying features. The right-quick rundown will *not* be all-encompassing of all features, nor is it intended to be used as the only details for identifying a mushroom. The rundown's purpose is to be able to get your eyes on speedy information.

This guide is principally for the eastern United States. This doesn't mean that these mushrooms don't grow elsewhere, but the seasonality and terrain may differ.

Mushrooms can be found growing in the previous season or into the earlier part of the next. I have categorized each mushroom into the season in which I find them in the most abundance in my neck of the woods (eastern Kentucky).

I have put the common size ranges for each mushroom in the "looks" rundown. Mushrooms play by their own rules, so take this with a grain of salt. Some mushrooms can get *massive* or be microscopic, depending on which maturation state you find them fruiting in. So, please be aware they *can* be smaller or larger than discussed here.

When it comes to the habitat of mushrooms, I've found certain mushrooms growing around different trees or terrains than they are commonly known to. I've seen mushrooms that are known to grow solely from wood appear to be growing from the soil. This usually means there is buried wood you're not seeing, and they are fruiting from it beneath the soil's surface.

Lastly, this is *not* an exhaustive list of all the edible mushrooms you can find in the great outdoors. In this book, I cover the basic and best edibles, in my opinion, and once you get these down pat, you can choose if you wanna branch out and explore even more.

Alright, let's get to it!

LEFT | Cluster of oyster mushrooms (*Pleurotus ostreatus*) on a dying ash tree.

CHAPTER 5

Edible Mushrooms of the Spring

Ain't spring such a magical time of the year? It's that oh-so-delightful transition from the colder months into a season where everything is coming back to life. Momma Nature wakes up some of the best wild edibles, and the land around us becomes revived and rejuvenated. Spring peepers start to sing their sweet song to remind us of what is just within reach—sweet springtime! The woods get greener, the days get longer, and honestly, my attitude gets adjusted. Spring kicks off a new mushroom huntin' season always worth being giddy for.

There are some iconic spring mushrooms I hunt for during this season. However, when people ask me which season I find the least variety of edible mushrooms in, I always say spring. That tends to shock people because they expect me to say winter, but that just ain't true for me. This spring chapter includes the fewest mushroom profiles of all the four seasons in this book.

Now, spring havin' fewer fungi offerings is not all bad. Remember, I said don't get discouraged! The spring is an awesome time of year to find some real heavy hitters, like morels. You can also find pheasant back mushrooms by the truckloads. It's a great time to scout out and familiarize yourself with ideal areas for the summer mushrooms that will be makin' an appearance before you know it. Plus, it's gonna be beautiful outside!

Scan here to get a more in-depth look at pheasant back mushrooms!

LEFT | If the trilliums are bloomin', it's time to be on the lookout for those springtime mushrooms.

MORELS (GENUS *MORCHELLA*)

Here he is! My first little guy I ever found. Words cannot describe the joy and the rush of finding your first morel. I highly recommend it if ya ain't done it yet.

Scan here to see me find my very first ever morel mushroom!

What better place to start the mushroom meat and taters portion of this book than with the highly sought after and elusive morel mushroom?! This feller is hillbilly gold right here. When the term "mushroom hunting" was coined, I believe they must have been mostly talking about these guys. Morels can sure be tricky to find, but when you do score your first one, there ain't nothin' sweeter. I'll always remember my first morel. It was a teeny tiny thing, but to me it was 10 foot (3 m) tall.

Even if someone doesn't know much about wild mushrooms, they will probably know what a morel is. They go by many a name: dry land fish, molly moochers, haystacks, hickory chickens, merkles, and they can come in a rainbow of colors referred to as blacks, grays, yellows, creams, and blondes. No matter what you call them, one thing that is universal is that they are dang delicious, buddy.

These mushrooms fruit only for a few weeks in the spring and get us hillfolk chompin' at the bit, ready to venture out into the woods. I've seen them pop as early as March and as late as May. It really depends on your weather that spring. Morels love it when the spring nights start to get warmer. They also love rain. A nice rule of thumb is to do your mushroom hunting a couple days after a good rain. Two important factors to take into consideration when you're about to prowl for morels would be the air temperature and the soil temperature. Once the spring nights start to hit 50°F (10°C) regularly, morels start coming out to play. They like a soil temperature of about 45°F to 50°F (7°C–10°C). Another indicator I use for morel hunting is a plant that goes by the name of mayapple (*Podophyllum peltatum*). It's an umbrella-shaped plant that grows in gobs. You can see them springing up on shady hillsides in spring, and when I see these plants emerging, I take it as my cue to get out there and look for morels.

In the early morel season, morels can be found chiefly at lower elevations and fruiting on the side of the hills that get warmed first by the sunshine (south- and west-facing slopes). As the season progresses, they start growing at higher elevations and more so on north- and east-facing slopes. I've also had good luck hunting morels in flood plains and burn sites.

Go Forth and Forage

A hefty yellow morel (*Morchella americana*) is always a sight for sore eyes in the springtime. Some folks call 'em blondes.

Now, as I mentioned before, these mushrooms can be elusive, and I've talked with many folks who have said they've looked for years and never found them. I've been there! Another piece of advice I give is when you morel hunt, cover *a lot* of ground. People can sometimes get too caught up in walking slowly and checking every little spot around them. I get it! You're afraid of missing one or walking right by it. However, I realized quickly that I was wasting precious time being too thorough and meticulous when I'd hunt for morels this way. I now prefer walking at a steady pace and keeping my head on a swivel, paying attention to trees or terrains morels seem to like, but I always keep on moving. Once I spot a morel, *then* I stop. Once the treasure has been located, that's when I start to move slower and be more observant. This practice has served me very well! Morels are also excellent at hiding. Some morels can blend right in with leaves and surrounding vegetation, especially when they are in their more immature state of growth. Fortunately, morels have a super unique appearance and texture. They are often conelike in shape and have a bunch of ridges and pits; think of a holey sponge. Keep your eyes peeled for this pattern amongst the forest floors. Once you see one morel, that visual pattern will easily register and sear itself into your brain, making it easier to spot the next ones. This goes for any mushroom you may be hunting for; once you see one, you'll start to see more. You'll be an eagle eye in no time if you put in the footwork.

There are a few different species of morels that occur in the United States, but we shall focus on the two most commonly occurring ones: the common or yellow morel and the half-free morel.

Scan here to peep some hefty half-frees!

Here's a bounty of gorgeous yellow morels. When you find one, there's usually more!

COMMON MOREL, YELLOW MOREL
(MORCHELLA AMERICANA, FORMERLY M. ESCULENTA)

I mostly find my morels growing solitarily, but sometimes you'll get lucky and find 'em growin' more tightly packed together.

LOOKS:

Common morels typically grow to be 3 to 6 inches (8 to 15 cm) tall, but I've seen them as big as a 2-liter bottle of pop. They sport a wrinkly, honeycomb appearance with many pits and ridges. They start out blackish to gray and become more yellow with age. The flesh is thin and fragile. The stem can range from white to yellow. The entire mushroom, from stem to cap, is hollow throughout.

STOMPING GROUNDS:

They prefer to grow directly from the ground around dead/dying hardwood trees, especially elm, sycamore, ash, hickory, poplar, and apple. Morels enjoy moist environments, so pay attention to creeks and riverbanks.

GROWTH PATTERNS:

Common morels can be found growing solitarily, in a scattered pattern, or sometimes in small clusters. I mostly see them growing solitarily.

*It is not recommended to consume morels raw or with alcoholic beverages.

SPORE PRINT COLOR:

light cream to yellow

FLAVOR NOTES:

earthy, nutty, meaty

POTENTIAL LOOKALIKES:

Some *Gyromitra* species are referred to as *false morels*. A true morel will be completely hollow throughout, whereas a false morel will not. *Gyromitra* also have more of a lumpy, brainlike appearance to their caps and they lack the holes a true morel cap has.

Gyromitra — a morel lookalike. **Edibility is debatable and not recommended.**

RIGHT-QUICK RUNDOWN:

- honeycomb cap with pits and ridges
- completely hollow from top to bottom
- cap attaches directly to stem

FRIED MORELS

INGREDIENTS

- 1 pound (455 g) clean morel mushrooms, cut in half lengthwise
- 1 cup (125 g) all-purpose flour, seasoned with salt and pepper to taste
- Enough oil in a cast iron skillet to fry in (I use vegetable oil, but you can use whatever you like)
- Extra salt on standby to season the tops

DIRECTIONS

1. Take the halved morels and dredge them in the seasoned flour until all are coated nicely, making sure you pay attention to get the flour into those nooks and crannies.
2. Pour enough oil into the cast iron skillet to deep fry and heat it up on medium heat. (Hot tip: Put the butt end of a wooden spoon into your oil. If it bubbles, you're ready to fry!)
3. Carefully place the dredged morels into the hot oil one by one and be sure to not overcrowd the skillet.
4. Fry, flip around with tongs, and babysit 'em until they're golden brown.
5. Remove the morels from the oil, carefully shaking off the excess grease, and place on a paper towel–lined baking sheet, butt-crack–side down to catch any extra grease.
6. Give a little sprinkle of salt to the top of those hot, crispy babies.
7. Place the baking sheet in the oven on the lowest heat setting to keep them warm as you fry subsequent batches.
8. After you've completely fried your haul, serve and eat immediately with a dipping sauce of your choice. Or you could just eat them straight up. Dipping sauce is not required with these flavor bombs.

Scan here to watch me fry up some morels!

Scan here to see some morel madness!

TO SOAK OR NOT TO SOAK

A point of contention amongst the morel hunting world is: to soak or not to soak? I am on the team that *never* soaks any of my wild mushrooms. I personally don't like what it does to the texture. Others swear by soaking their morels in salt water for flavor and to coax out any little buggers that may have set up shop in the mushroom. If I find a morel infested with bugs, I usually leave it be. Bugs gwotta eat, too. Also, if you leave a buggy mushroom behind, it is then left to the forest to spread its spores and love to further encourage mushroom growth. For the morels I harvest, I simply do a quick rinse under running water in my sink for the least amount of time as possible, while gently wiping away any dirt and debris. Mushrooms are like a sponge and can quickly become waterlogged. Morels often have a thin, brittle texture, so I do all I can to prevent them from getting soggy and falling apart. I want to preserve that beautiful morel integrity! You choose your own adventure, though! As long as you're having fun and getting a full belly, that's all that matters.

Edible Mushrooms of the Spring

EASTERN HALF-FREE MOREL (MORCHELLA PUNCTIPES)

Half-free morels are muuuch easier to spot pokin' up out of leaf litter than their sneakier cousins, common morels.

In morel season, you're likely to find half-free morels popping up first. It's like the opening act for the common morel that will hit the main stage later in the spring. I've found it's always good to check out the opening act, even if the main act is what you're most interested in. You might just realize the opener is better than you were expecting. Same thing for morels. Work with me here. So, if you haven't found common morels yet in the springtime, but you've found yourself a half-free morel, rest easy, because the big dogs will be out soon and growing in similar places!

Half-frees are muuuch easier to spot than their common morel cousin. Half-frees stand tall and proud and have a light-colored stem that sticks out amongst the leaf litter. When I found my first half-free morel, I had been out all day looking for common morels. I would ride Wanda (my 1992 Honda FourTrax 300 four-wheeler) to places that checked all the morel habitat boxes. I'd get off, cover lots of ground searching all over for 'em, but no dice. I had finally given up on the mission. Then, as Wanda and I were leaving, I had to stop and open/close a gate to get back to the road to head home. I looked down and by golly, right there by the trail was a big, chonky half-free morel! I couldn't believe it. I let out my iconic mushroom shriek that shook trailer winders six hollers over. All that time walking in the woods and then there's a morel, granted a half-free, but right there by the daggone trail! I kept creeping along that trailside and found several tall, beautiful half-frees. I wasn't skunked after all!

Another mushroom that comes to mind for me as a potential lookalike for the half-free morel is the oh-so-lovely stinkhorn mushroom. Remember him? He's that's stinky, slimy feller that pops up in your flower bed and reeks to high heaven. A few different species of stinkhorns resemble the morel, but I think that *Phallus hadriani* and *Phallus impudicus* take the cake as far as likeness goes. Their long stem with a small, ridged cap could maaaybe fool ya into thinkin' it's a half-free, but to tell the difference, do one simple thing—sniff it. If the cap has goop on it and smells like someone took a crap on a dead body, it's a stinkhorn and you definitely don't want it. Stinkhorns start out as a little egg in their immature form and some folks *do* eat stinkhorn "eggs" raw, but it ain't for me, babe. I'll stick with my morels, thank you.

A leanin' long and brittle boy shows himself. Note his smaller cap and longer stem as compared to the common morel.

LOOKS:

Half-free morels typically grow to be anywhere from 2 to 8 inches (5 to 20 cm) in height, with the head of the mushroom being ½ inch to 2 inches (13 mm to 5 cm). The head is brownish to black in color and exhibits that iconic morel honeycomb pattern with pits and ridges. It has a more tapered tip than the common morel. The stalk is lighter in color, ranging from white to off-white to yellowish. The stalk is attached up into the head of the mushroom, leaving the lower half margin of the head hanging free (hence half-free) from the stalk. They are hollow throughout. They are much more delicate and brittle than common morels. They crumble very easily, so pack them with care if you're gonna put them in your foraging sack. I have been known to cradle my haul gently in my arms and carry them home like the tender babies they are.

STOMPING GROUNDS:

They prefer to fruit around hardwood trees in the early spring. Half-free and common morels love the same habitats, so think elm, sycamore, ash, hickory, poplar, and apple trees.

GROWTH PATTERNS:

They can be found growing singularly or in scattered groups directly from the ground. I've found them growing in nearly a single file line with a handful of steps between each mushroom.

SPORE PRINT COLOR:

light cream to yellow

FLAVOR NOTES:

They are milder in taste than the common morel, but still lend a rich, nutty flavor.

*I like to mix my half-free morels with other common morels to beef up the flavor and texture that half-frees can lack.

POTENTIAL LOOKALIKES:

Verpa bohemica may resemble half-frees due to their small, wrinkly, thimblelike caps. You can distinguish these from half-free morels because *Verpa bohemica* has more of what are like folds or wrinkles on its cap versus the honeycomb pattern with pits as seen in the half-free morel. Another big difference between *Verpa* and morels is that *Verpa* mushrooms have a cap that attaches only at the very tippy top of the stem. It looks as if someone just delicately placed the cap atop the stem, kinda like a lampshade. Some folks consider the edibility of *Verpa bohemica* to be debatable, but it seems most people cook 'em up and eat 'em often and have lived to tell the tale.

RIGHT-QUICK RUNDOWN:

- elongated stem
- smaller, wrinkly, honeycomb cap
- bottom portion of cap hangs freely from stem
- completely hollow from top to bottom

PHEASANT BACK (CERIOPORUS SQUAMOSUS)

Nope. Your eyes ain't failin' ya, pal. These are two giant pheasant back mushrooms I scored off a loaded boxelder tree. I sautéed the tender outer edges up and topped some yummy burgers with them, and I dehydrated and powdered the rest to create some of my mushroom seasonin' blends. It all got used!

Ahhh, the always reliable pheasant back mushroom, or as some call it, *Dryad's saddle*. This fungus is one that is readily available in its seasons. Yes, I said it as plural, *seasons*. Pheasant back first comes out to play in the spring and then makes a much welcome comeback in the fall. I decided to place this mushroom here in the spring section so you can get a head start on its first flushin'. You can harvest until the cows come home on this one because it is super plentiful and super easy to ID. If you see one, look for more on that same tree or on surrounding trees. I once found a boxelder tree with thirty-two separate pheasant back mushrooms on it. They were growin' pretty high up on the tree, but luckily, I carry a pole saw in my vehicle for such challenges. I cut me down a few of them suckers with ease. Consider travelin' with a pole saw.

It is no mystery how this mushroom was assigned its common name; it looks like the rear end of a pheasant. The mushroom cap is adorned with an iconic, beautiful feathering pattern. These puppies sure come in a range of sizes. I have found cute little baby button pheasant backs, and I have also found pheasant backs as big as a car hood—and that is not an exaggeration at all. These guys get heckin' hefty.

Although pheasant backs give a beautiful showin', they are not the greatest culinary mushroom out there. Folks are purdy split on its flavor/texture profile. With a strong cucumbery flavor that can easily overpower a dish, ya gotta be crafty when workin' with pheasant backs in the kitchen. I like to thinly slice them and sauté or roast them with punchier ingredients, that way my food doesn't taste only of the pheasant back. I think it's a nice addition to stir-fry or stews. This mushroom also has a tendency to be tough and chewy, but there are a few ways around this. Firstly, you can opt to pick only small, young specimens of this mushroom because they will be nice and tender throughout. If you come across a pheasant back that is large and in charge, harvest only the tender outer edges of the mushroom so it won't be like chewin' on boot leather. Your knife should easily slide and glide through the mushroom's flesh at the tender parts, serving as a good way to know you're doing it right. If your knife is met with any resistance, it's liable to be too thick and rubbery to eat on that part of the cap. The mushroom gets tougher as you work your way closer to the stem, so those parts ain't great for cooking. What I do instead with these tougher, chewy parts is dehydrate them and grind them into powder to flavor food with later, or I use the leftovers to make a flavorful stock. Waste not, want not, y'all!

The pheasant back is a polypore mushroom, meaning it has small holes, or pores, under its cap. It is a tougher textured mushroom that grows from wood.

LOOKS:

The cap typically grows up to 18 inches (46 cm) wide, but as I mentioned earlier, they can get an awful lot bigger than this average growth estimate. The cap begins as a small circle and becomes larger and kidney- or fan-shaped as it matures. The cap has creamy white tones as the base color, overlaid with brown to black feathery scalelike patterns, reminiscent of a pheasant's tail markings. The underside of the pheasant back should have honeycomb-like patterned pores. The pores become larger and more open with age. This mushroom has a thick, sturdy, but stubby rounded stem growing up to 3 to 4 inches (7.5 to 10 cm) long. The stem connects directly to the wood or tree it is growing from. The stem is whiter when young and darkens to black with age. The mushroom is thicker at its center and thins out toward the cap margins.

STOMPING GROUNDS:

Pheasant backs prefer to grow directly on dead/dying hardwood trees, especially maple, elm, and boxelder. I find most of mine on trees that are near creeks and riverbanks.

GROWTH PATTERNS:

They can be found growing by their lonesome, but more often than not, they are found growing gregariously on the same trees, sometimes in giant overlapping clusters.

SPORE PRINT COLOR:

white

FLAVOR NOTES:

cucumber/watermelon with a hint of nuttiness

POTENTIAL LOOKALIKES:

The train wrecker mushroom (*Neolentinus lepideus*) can somewhat mimic the cap of a pheasant back mushroom. The train wrecker has gills on the underside instead of pores, so telling the difference is a no-brainer.

RIGHT-QUICK RUNDOWN:

- brown to blackish feathery pattern (scales) on cap
- underside has pores
- stubby and thick off-center stem
- growing directly from wood
- smells like cucumber/watermelon

Edible Mushrooms of the Spring

WINE CAP (STROPHARIA RUGOSOANNULATA)

Wine caps are a welcome sight in the springtime. They are indicative of longer, sunnier days ahead, and they get me pumped right up for other highly anticipated warmer weather mushrooms! Wine caps can fruit all the way from spring up into the fall season, but I find them more in the spring. All parts of this bad boy are edible, and the texture is nice and firm, makin' it a good, hefty addition to your dishes.

Wine caps are what I refer to as a more urbanish mushroom, meaning you can find them in less rural, woodsy areas. They are often seen sproutin' right from the grass in yards or even out of wood chips and mulch in flower or garden beds. With that being said, when harvesting this mushroom, please be sure the area in which you collect them from has not been sprayed or treated with any sort of chemicals. I prefer my mushrooms without a side of herbicide.

I have found wine caps out in the more rugged wilderness, too. They tend to grow from woody debris in moist areas. I've had good luck spottin' 'em near creeks and rivers. I've noticed in my adventures that I often find wine cap mushrooms and pheasant back mushrooms around the same time and in the same areas. So, if ya see a pheasant back, check the surrounding grounds for wine caps, and vice versa.

Wine caps are a mushroom that is easily cultivated. I have had a wine cap garden bed for a few years now, and it's super easy to manage. Purchase wine cap spawn and spread it into hardwood chips and straw to grow your own little patch to enjoy at home.

This peek under the wine cap's hood shows ya the features it is known for: grayish-purplish gills packed closed together and attaching to the stem, a funky, toothy skirt, and a lighter off-white to yellowish stem.

LOOKS:

Wine caps typically grow to be up to 6 to 8 inches (15 to 20 cm) tall, with the cap ranging anywhere from 2 to 8 inches (5 to 20 cm) wide, depending on the level of maturation. I have found *enormous* wine cap mushrooms before, so remember, they can indeed grow to be much larger. The smooth cap starts out just as its namesake suggests: as a purdy, deep-red wine color. The cap turns more light brown and becomes duller in color with age. The cap begins as a rounded, bell shape and then flattens out as the mushroom matures. The gills are gray to purple to brownish in color and are spaced closely together. The gills attach to the stem. The stem is white to off-white in color, eventually yellowing, and can have a slightly bulbous base. This mushroom starts with a partial veil covering its gills when young. As it grows, the veil breaks away, exposing the gills and leaving behind a spikey/toothy ring on the stem.

STOMPING GROUNDS:

Wine caps like to grow from the ground in woody debris, mulch, lawns, and in moist areas.

GROWTH PATTERNS:

They can be found growing singularly, in a scattered pattern, or in small clusters.

SPORE PRINT COLOR:

purplish brown

FLAVOR NOTES:

mild, earthy, with hints of raw potato

POTENTIAL LOOKALIKES:

Wine caps can kinda sorta resemble some *Agaricus* species of mushrooms. The way to tell the difference is: *Agaricus* mushrooms often have pinkish-colored gills when they are prime but the gills turn darker brown with age. *Agaricus* mushrooms also have a darker brown spore print. The *Agaricus* mushroom genus has both edible and poisonous species, so as always, don't put it in your mouth if ya ain't sure about it.

RIGHT-QUICK RUNDOWN:

- burgundy cap
- whiteish to off-white stem
- grayish-purplish gills
- spiky, toothy ring on stem
- likes to grow from wood chips

WOOD EAR (AURICULARIA ANGIOSPERMARUM)

Wood ears, or as some call them, jelly ears, sit pretty amongst the jelly fungi family. He's wiggly, jiggly, and fun to hold up to the side of your head and holler, "HUH?! WHAT'D YA SAY?!" Whoever came up with the common name of this mushroom is really hittin' the nail on the head because this mushroom grows from wood and looks like an ear—hence the name, wood ear.

These fellers can grow throughout multiple seasons, startin' in the spring and explodin' again in the mid-fall through the early winter (buuut, of course, I've found them in the summer months, too). I've come across trees plum-covered in wood ear well into the winter topped with blankets of snow. Depending on the kind of weather you spot them in, wood ears can have varying textures and colors. In their true, Mother Nature–given form, they are wobbly with a rubbery give to 'em and are brownish in color. If there's recently been a frog-strangler and flat poured the rain, wood ears become more gelatinous and can swell up full as a tick. If it's bone dry and hotter than blue blazes outside, you can also find 'em all dried up, crunchy, and darker brown to black in color. Despite the shriveled-up, crusty appearance of a wood ear in times of little rainfall, they can still be foraged! Take those puppies home and stick 'em in a sink full of water and wait for the mushroom magic. You can watch them swell up to their original form right before your eyes.

These wood ears are showin' off their fine, fuzzy white hairs and the lobey folds they are so appropriately named for.

LOOKS:

Wood ears typically grow 1 to 6 inches (2.5 to 15 cm) wide and are cup-shaped and earlike. They range from brownish red to a light to medium brown in color. They do not have a stem and attach directly to the wood they are fruiting from. The cap is smooth in texture, sometimes with tiny white hairs that can give it a frosty appearance. Wood ears often have wrinkles, lobes, and veinlike lines running through them. They are rubbery and should easily move around, fold, and bend when handled.

Wood ears love to stick close with their buddies and grow directly from dead/dying hardwood.

STOMPING GROUNDS:

They prefer to grow from the wood of dead/dying hardwood trees, including on stumps, fallen logs, branches, and limbs. I find mine especially on boxelder trees near water.

GROWTH PATTERNS:

Wood ears can most often be found growing in groups, colonies, or dense clusters on wood. They can also be found growing by their lonesome or scattered on wood, but this occurs less often as they like the close company of their buddies.

SPORE PRINT COLOR:

white

FLAVOR NOTES:

Wood ear does not have a ton of flavor and takes on the flavor of whatever sauce/broth/seasoning you cook it in. Some folks describe the flavor as slightly earthy and mild. Wood ear is mostly used for a textural component in a dish. Wood ear is especially popular in Asian cuisine.

POTENTIAL LOOKALIKES:

Amber jelly roll (*Exidia recisa*) is a common lookalike for the wood ear mushroom. Amber jelly roll is different in that it is usually darker brown in color, smaller in shape, and is squishier and more bloblike. You can squeeze it and easily change the shape or stretch it. Wood ear is firmer, not as easily stretched, and its shape cannot be shifted around like amber jelly roll. Amber jelly roll does not have tiny hairs and is typically wrinklier than wood ear. Amber jelly roll likes to grow more on branches and sticks instead of on the larger pieces of wood that wood ear prefers.

RIGHT-QUICK RUNDOWN:

- brownish red to light to medium brown in color
- ear-shaped
- rubbery, jellylike texture
- veiny markings
- growing from wood

DEER MUSHROOM (*PLUTEUS CERVINUS*)

Deer mushrooms have white, pinkish-tinted, crowded gills that are free from the stem.

Deer mushrooms often have a pale, brownish cap with a center "bump" that is darker in color.

Deer mushrooms are downright common in the spring and can continue to grow into the summer and fall months. And yep, deer *do* like to munch on these babies, as well as other woodland critters. You're likely to find them all over the dang place, springin' up from logs in the beginning of spring at a time when there are not many other edible mushroom options. This is why I love the deer mushroom; they are reliable and bountiful. Deer mushrooms aren't appreciated enough in my opinion. Some foragers will skip on 'em entirely. As for me, I'll take any and all edible mushrooms and work with 'em for what they are. They're all special in their own right. Sure, the taste is not anything that is gonna make your tongue slap your brains out, but I take this as a challenge to turn them into somethin' tasty! A "con" some folks may put this mushroom down for is that they're not very meaty. It's purdy much a whole lotta gills and a little bit of cap meat. I say to those people: There ain't nothing wrong with throwing some sliced up deer mushrooms into any dish you'd use store-bought button mushrooms in. The flavor is similar and mild. The deer mushrooms just won't provide as much bulk—and that's okay. Supplement with other beefier mushrooms if it's the girth you crave.

Deer mushrooms are one of those mushrooms that aren't very stout. They are fragile and brittle with a tendency to crumble in your foragin' sack as you romp through the woods. When harvesting these mushrooms, make sure you are storing them in a protected way and not letting them bang and clang against each other. This ensures you get your deer mushrooms home in one piece.

The bugs looove a good deer mushroom. Oftentimes when ya flip the mushroom over to get a peek at its gills, you'll find them sprinkled with tiny insect glitter. This is normal, and I usually just blow the bugs right off. If the mushroom is crazy infested and the bugs are not easily brushed or blown off, I pass on it. Deer mushrooms also can get a little floppy and sad in high heat and heavy rains. I recommend pickin' only fresh, healthy mushrooms so you don't end up with a soggy mess in your bag.

Deer mushrooms love to shoot up in groups from dead/dying wood. Also, it looks like a critter may have nibbled itself a lil' samplin'.

LOOKS:

Deer mushrooms typically grow to have both a cap width and stem length of 5 inches (13 cm) (but they can grow bigger). The cap is pale to darkish brown in color, with the center of the mushroom having a circular raised bump that is darker in color, called an *umbo*. The cap flattens as it ages. It can also crack and separate at the cap margins with age. The gills are packed tightly and begin as white and shift to a pinkish or salmon color soon after. The stem is white to off-white in color and can either have a smooth, bald surface or have grayish brown fibers on it.

STOMPING GROUNDS:

Deer mushrooms prefer to grow in hardwood forests, but aren't picky, sometimes fruitin' near conifer trees. They grow from dead or rotting wood. They can appear to be growing from the ground at times, but it is actually fruiting from buried, decaying wood hidden away beneath the surface.

GROWTH PATTERNS:

They can be found growing singularly, scattered, and in small groupings from wood (or buried wood).

SPORE PRINT COLOR:

brownish pink to salmon pink

FLAVOR NOTES:

mildy earthy with hints of radish

POTENTIAL LOOKALIKES:

The platterful mushroom (*Megacollybia rodmanii*) is a potential lookalike. However, the platterful mushroom cap grows to be much larger than the deer mushroom and has white to off-white gills with wider spacing, with the gills broadly or narrowly attaching to the stem. I do not recommend consuming platterful mushrooms.

*Platterful mushroom (*Megacollybia rodmanii*) is a potential lookalike for deer mushrooms.*
*Edibility debated

RIGHT-QUICK RUNDOWN:

- pale to dark-brown cap, sometimes with a darker color bump in the center of the cap
- white to pinkish gills
- gills smell slightly of radish
- growing from wood (or buried wood)

CHAPTER 6

Edible Mushrooms of the Summer

Buckle up, buddy. This is the one. This is the season where you're gonna start to notice mushrooms at every single turn. The forest floor and the trees are gonna be dynamic and adorned with mushrooms of so many colors and sizes and shapes and textures and patterns, and it's so good for the soul—and of course, the tummy, too.

The summer is when I rake in my biggest hauls of the year, and it's when I find the most variety. I love to load up on good ol' summertime mushrooms and preserve them to enjoy throughout the year. Each and every day of this season is an opportunity to frolic your fanny off and find top-choice edible mushrooms.

Now, let's see what's out there for grabs during the bountiful and benevolent summer season.

Scan here to check out some summertime mushroom huntin'!

LEFT | When the summer greenery is full, so can be your mushroom huntin' sack!

CHANTERELLES (GENUS *CANTHARELLUS*)

A good place to kick this thing off is with the oh-so-coveted and admirable chanterelle mushroom. There are many types of chanties that fruit all over the world, and my little corner of Appalachia happens to house plenty good ones. In the following profiles, we take a look at some of my favorites, including the golden, smooth, peach, Appalachian, and cinnabar chanterelles.

Before we jump in, let's go over some chantie basics since we're fixin' to look at a whole rainbow of 'em in the pages ahead. Many chanterelles carry a package deal of some key identifying features across their genus. Firstly, chanterelles are trumpeted or funnel-shaped, often with wavy cap margins. In their younger stages, they *can* be more buttonlike in appearance and lack the wavy margins. As they mature, they eventually develop their iconic trumpeted shape and curly edges if left alone to complete their glow up. Chanterelles (with the exception of smooth chanterelles) have those false gills we talked about in chapter 2. Instead of feathery, fragile gills, chanterelles have raised ridges and folds under their cap that are not easily broken or moved around. These false gills of the chanterelle are also forked, meaning they branch off and prong, especially near the cap margin. The false, forked gills are also decurrent, meaning the gills continue to run down the upper portion of the stem. You will usually be able to see the gills from the side of the mushroom without having to flip it over when the chanterelles are prime.

Scan here to watch me explain the differences between chanterelles and jack-o'-lanterns!

GOLDEN CHANTERELLE
(*CANTHARELLUS* SP.)

Previously referred to as *Cantharellus cibarius*, which is now known to be a European species, the beautiful, beamin' golden chanterelle mushrooms signify the beginning of summer for me every year. The first one I spot of the season always sends me into a feral fit of high kicks, hootin', and hollerin'. Chanterelles are a super easy mushroom to hunt for. You're looking for vibrant pops of golden oranges and yellows among the hills in contrast to the leaf litter, soil, and mosses. They purdy much stick out like a sore thumb. Another added bonus: Chanterelles love the company of their brethren, so if you see one, I'd bet the farm there's more.

This lovely couple of golden chanterelles display all the right stuff: trumpeted shape, wavy margins, and decurrent false gills.

The golden chanterelle loves to start sayin' howdy when the days get real hot and real humid. They really like to pop up after heavy rains. My best months in eastern Kentucky for finding these guys are usually July and August, but of course they can be found before or after this window as well, dependent on the weather. If you live in more southern parts of Appalachia, you're likely get them earlier as y'all get the hotter days before I do. If you're further north, your season may come a little later. You mainly want to pay attention to the weather patterns where you're located to get an idea of when it's best to hit the trails. Chanterelles like it hot and wet!

Chanterelles form mycorrhizal relationships with trees, and most guides describe them as growing around both hardwoods and conifers. From my own personal treks, I have the best luck with hardwoods. There's a holy trinity I seek out when I'm huntin' chanterelles that I have turned into a mnemonic device of sorts. Work with me here. So, chanterelles are the bomb, baby. The B-O-M-B. And I find chanterelles mainly around **b**eech, **o**ak, and **m**oss, **b**aby. Get it? Like it? Use it if so. Chanterelles have also been known to fruit happily alongside spruce and hemlock trees. Basically, just get out in any kind of woods in the summer, look around for bright orange and yellow, and you're bound to find them if you're willing to put in the footwork.

Another trick I like to employ when foraging is lookin' for other species that are known to occur with my target mushroom. I like to look for something specific when I'm hunting for chanterelles and that is the cutie mushroom called the jelly baby (*Leotia lubrica*). These serve as a good indicator that chanterelles are likely to be in the area. I follow 'em like a trail of breadcrumbs to big honkin' chanterelle patches all the time. Jelly babies like the same stomping grounds as the chanterelle.

They're a jelly club fungus that are often yellowish in color, itty bitty, and rubbery. They like to grow in tight clusters from the forest floor. If you're lucky, you just might get to see jelly babies that have developed a gorgeously colored deep, dark-teal top. This is just a chanterelle-hunting trick I learned and wanted to share with ya to hopefully make your hunts more fruitful. Peep at the variations of jelly babies in the photos on this page.

Chanterelles are one of the most prized wild mushrooms folks hunt for, and for good reason. They are delicious! The flavor and texture of these mushrooms make any savory dish a better one. The texture is phenomenal. They have a firm, meaty texture, while still having a nice tender bite to 'em. The flavor is so interesting and downright addictive. They have an earthy, nutty taste, but then on the back end comes this nice fruity note (often compared to apricot or peach without the sweetness) and a hint of pepperiness. They'll make ya lick your eyebrows clean off your forehead.

This perfect patch of jelly babies led me to chanterelles last summer. Jelly babies are a super-duper easy way to track down some chanties!

This particular jelly baby led me to some cinnabar chanterelles! You can see the variation of the top here, sportin' that gorgeous teal color.

LOOKS:

Chanterelles typically grow to be around 4 inches (10 cm) tall, with the cap stretching up to 5 inches (13 cm) wide. The cap is smooth and can range from pale yellow in color to vibrant yellowish orange. They are often funnel-shaped with a depressed center, having thin, wavy cap margins. Under the cap, the gills will be false, forked, and decurrent. The stem is usually off-whiteish or similar in color to the cap, tapering slightly downward in shape toward the ground.

STOMPING GROUNDS:

They prefer to grow directly from the soil in both hardwood and coniferous forests. I find them especially amongst beech, oak, and moss, baby (BOMB).

GROWTH PATTERNS:

They can be found growing singularly, in scattered groups, or gregariously.

SPORE PRINT COLOR:

white to pinkish to yellow

FLAVOR NOTES:

earthy, nutty, slightly fruity, peppery

POTENTIAL LOOKALIKES:

Chanterelles have a few naughty imposters that tend to grow in the same areas at the same time of year. The most common chanterelle lookalike would probably have to be the jack-o'-lantern mushroom (*Omphalotus illudens*). This mushroom is *not* edible and will make you really sick if consumed. Here's the way to tell you have a nasty jack instead of a tasty chanterelle. Jacks have true gills, are brighter orange in color, grow larger, grow in more dense clusters, and are not white in the middle. Jacks also like to grow from wood or buried wood and chanterelles do not.

Another potential lookalike is called the false chanterelle (*Hygrophoropsis aurantiaca*). Go figure. (They just don't make it that easy on us, do they?) So, there are some key differences in the real deal and the false feller. The main way I tell the difference is all in the gills. The false chanterelles have more narrow, super forked, crowded gills that are easily separated, versus the chanterelle which has the more sturdy false gills that are not as crowded and not as forked. Chanterelles also are much firmer in texture and have a slightly fruity scent to 'em. Edibility is debatable on false chanterelles. Some eat them, some report GI issues, so ya know what I'll say: I don't recommend putting 'em in your mouth with all this conflicting info.

Aaaand another guy that might fool ya is the woolly chanterelle (*Turbinellus floccosus*). He can become one biiiig dude. The woolly chanterelle has more of a pronounced vase shape, often lookin' like you could just pick it up and drink from it like a medieval goblet. It has an off-white outer surface with wrinkles or veins that

Continued on next page

Brightly colored jack-o'-lanterns can trick ya into thinkin' you might have some big ol' chanterelles from a distance. Once you get closer, you'll see that jack-o'-lanterns get much larger in size, grow in dense clusters, and thrive on wood, especially lovin' the base of dead trees. *****Not edible**

Golden Chanterelle, con't from previous page

The false chanterelle differs in its gill makeup. Note the gills that are closely crowded, more forked, and lighter in texture. *Edibility debated

run all the way down the stem. The main difference is the woolly chanterelle does not develop an obvious cap like the golden chanterelle. The stem is kinda hard to discern from the actual cap area as they appear to be connected as one whole structure. Research has shown that the woolly chanterelle is not as closely related to chanterelles as once thought, and it is actually more genetically similar to the genus *Ramaria* (coral mushrooms). Yet again, edibility is conflicting and not recommended by ya girl on this one.

With careful observation and comparison, you'll be able to tell the difference in these lookalikes before ya know it.

The woolly chanterelle does not have an obvious cap structure like the golden chanterelle. *Edibility debated

RIGHT-QUICK RUNDOWN:

- pale yellow to orange in color
- trumpet shape
- false gills
- forked gills
- decurrent gills
- white in the middle (pulling apart like string cheese)
- slight smell of apricot

Scan here for a delicious Chanterelle Cream Sauce Linguine recipe!

SMOOTH CHANTERELLE
(*CANTHARELLUS LATERITIUS*)

This chanterelle is *exactly* like the golden chanterelle we reviewed previously, except for one thing, it ain't got no gills! Well, for the most part, it's bald. The underside of a smooth chanterelle will be orangey to pale yellowish in color and be either totally smooth or have just a jag-bit of faint ridges. They have the same funneled shape, fruity smell, and sprout in similar habitats as golden chanterelles. Since all of the features and information are the same, minus it being a smooth customer, we won't have to go over all that again here. So, simply apply the same things from the previous chanterelle profile and you'll be golden—pun intended.

RIGHT-QUICK RUNDOWN:

- pale yellow to orange in color
- trumpet shape
- *no* gills or very faint ridge lines under the cap
- white in the middle (pulling apart like string cheese)
- slight smell of apricot

PEACH CHANTERELLE
(*CANTHARELLUS PERSICINUS*)

Yep, here's another'un. If the goldens and the smooths weren't enough for ya, there's also a purdy pinky peach chanterelle that's out there for grabs. Again, all the same rules apply here for this chanterelle as with the golden chanterelle, but this'un is just a different, fun color. It yields a whitish to pinkish spore print. I find the taste and texture to be very similar to the golden chanterelle and often find them fruiting together in hills. The peach chanterelle is a much rarer find than its golden cousin, so do a lil' dance when you spot one. They are kinda like me in that they seem to have a likin' for the Appalachian Mountains, and you won't find them very often anywhere else.

RIGHT-QUICK RUNDOWN:

- pink to peach in color
- trumpet shape
- false gills
- forked gills
- decurrent gills
- white in the middle (pulling apart like string cheese)
- slight smell of apricot

Edible Mushrooms of the Summer

APPALACHIAN CHANTERELLE
(CANTHARELLUS APPALACHIENSIS)

Now ya know this makes me proud as a peacock. There's actually an *Appalachian* chanterelle! Just like Appalachia, it's a little more unique, and it differs from your run-of-the-mill golden chanterelle in a few ways. Even though its common name is Appalachian chanterelle, it is not confined to growing only in Appalachia. It also grows in other parts of North America, too.

The main identifying feature of these chanterelles is the center of the cap. It has a dark patch on it when mature, whereas other chanterelles do not sport this feature. Appalachian chanterelles also vary in color, from yellowish brown to a yellowish orange. Appalachian chanterelles tend to be smaller and more delicate than beefier chanterelles in the genus. Don't get it twisted, thought, they are definitely a tasty edible, but you do have to find quite a few to make it worth your while. They also are often damp or wettish when younger, drying out more with age. So, a slighty soggy Appalachian chanterelle is totally normal and edible. It's gonna have all those iconic chanterelle features, too: funnel-shaped with false, forked, decurrent gills.

LOOKS:

Appalachian chanterelles typically grow to be ½ inch to 2 inches (13 mm to 5 cm) wide across their cap and 3 to 4 inches (7.5 to 10 cm) tall. They are funnel-shaped and start out with a curved downward cap initially, flattening with age, and developing a central depression, or a small dip, in the center of the cap, with the depression turning darker brown in color when mature. The cap ranges from yellowish brown to yellowish orange. The top of the cap is smooth, with the margins rolling downward at first, and then the margins get wavier and more turned upward with age. The underside has forked, false, decurrent gills. The stem is typically similar in color or slightly lighter than the color of the cap, becoming hollow with age. The flesh is thin, flimsy, and pale to yellowish throughout. These mushrooms can also sometimes appear to have a greenish tinge to them.

STOMPING GROUNDS:

They prefer to grow like other chanterelles because they are the B.O.M.B. (around beech, oak, and moss, baby!), fruiting directly from the soil or moss.

GROWTH PATTERNS:

They can be found growing singularly, scattered, or gregariously.

SPORE PRINT COLOR:

white to creamy white

FLAVOR NOTES:

earthy, nutty, slightly fruity, peppery

POTENTIAL LOOKALIKES:

See *Hydrophoropsis aurantiaca* (false chanterelle) discussed previously in the golden chanterelle profile.

RIGHT-QUICK RUNDOWN:

- yellowish brown to yellowish orange in color
- small
- funnel shape
- false gills
- forked gills
- decurrent gills
- depressed center with darker patch on center of cap
- thin flesh
- slight smell of apricot

Scan here to check out some summertime chanterelles!

Edible Mushrooms of the Summer

CINNABAR CHANTERELLE
(CANTHARELLUS CINNABARINUS)

Nope! We ain't done yet, buddy! Here's another chanterelle for ya—the cinnabar! Some call it a red chanterelle. This tiny feller gets his name from his beautiful cinnamon shade of bright red, reddish orange, to a pinky red. Due to its vibrant colors, it is easy to spot this mushroom. Like the Appalachian chanterelle, cinnabars are a smaller chanterelle, requiring quite a bounty to make a dish from. I usually will mix cinnabars in with my other summer chanterelle finds since they can be quite tiny. I've found that cinnabars don't necessarily have the fruity bouquet to 'em like other chanterelles, but they are still a choice edible and so fun to find.

LOOKS:

They typically grow to be as much as 2 inches (5 cm) wide and 2 inches (5 cm) tall (I usually find them to be smaller). They are often funnel-shaped, with thin flesh. Cinnabar chanterelles start out with a smooth, downward curved cap initially, that flattens and becomes wavier with age and develops a central depression, or a small dip, in the center of the cap. The cap color ranges from bright red to reddish orange to reddish pink. The underside is similar in color with false, forked, decurrent gills. The stem is similar in color (sometimes lighter at the bottom), small, thin, and usually tapers downward.

Cinnabars, small but mighty, really pop with their vibrant colors, makin' 'em an easy-to-hunt mushroom.

STOMPING GROUNDS:

They prefer to grow like other chanterelles because they are the B.O.M.B. (around beech, oak, and moss, baby)! I also often find cinnabars around the edges of creeks, right next to the water, in sandy soil.

GROWTH PATTERNS:

They can be found growing singularly, scattered, or gregariously.

SPORE PRINT COLOR:

pinkish cream

FLAVOR NOTES:

earthy, nutty, slightly peppery

POTENTIAL LOOKALIKES:

Some *Hygrocybe* and *Hygrophorus* mushrooms can resemble cinnabars. The easiest way to tell the difference is that *Hygrocybe* and *Hygrophorus* have true gills instead of false gills, and they have a waxy, slick flesh. Edibility is debatable throughout this genus, and some are downright inedible.

*This scarlet waxcap (*Hygrocybe coccinea*) is an example of a Hygrocybe that could resemble a cinnabar. Its waxy flesh and widely spaced true gills will clue you in that it's not the chanterelle you are after.* *Edibility debatable*

RIGHT-QUICK RUNDOWN:

- bright red to reddish orange to pinky red in color
- small
- funnel shape
- false gills
- forked gills
- decurrent gills
- thin flesh

BLACK TRUMPET (*CRATERELLUS FALLAX*)

Aw, heck! What's one more summer chanterelle family member? Black trumpets are one of my personal favorite chanterelles to hunt for and play around with in the kitchen. They are a dainty but wicked-looking mushroom. They can look like they're straight out of a Tim Burton movie, and they have that dark, gothic charm to them. The taste and texture make them extremely versatile and a welcome umami addition to your supper plate.

Black trumpets are one of those mushrooms that can be a little difficult to spot initially, but once you see one, you're liable to see so many more. Their dark appearance requires a sharper eye than more brightly colored mushrooms, but once you nail down their preferred habitats, you'll be able to find them growing in blankets in the late summer to fall. Here's a trick I began using while hunting for these guys: Once I spot one, I then hunker down and get eye level with the forest floor, looking outward in all directions from that black trumpet to catch sight of any others that are sticking up. They like the company of their buddies, so it's a dang good possibility you'll find more than just one after you spot the first.

If you aren't familiar with black trumpets and encounter one, you might think it's an old, dead, dried up mushroom, but it ain't! This is just how they look, and we love them for it. They have an older, worn vibe to them, so doing a quick feel of the mushroom will clue you in on its freshness. They should have a flimsy give when you give 'em a boop. They can become more brittle and drier if they've been hanging out too long in the sun. Your intuition is pretty reliable when evaluating the freshness of a mushroom; if it feels and smells nice, it probably is. If it stinks, has a questionable texture/feel to it, or is infested with bugs, it's probably not primo. Leave it be.

Another thing to note with black trumpet patches, when you are fortunate enough to stumble across one, is that the area can also be a utopia for ticks. I once found an enormous black trumpet patch and got so dang excited that I plopped right down on my hind end and got to pickin'. While cramming them into my sack, I soon noticed little black dots all over my arms. It was then I realized I got plum eat-up with tiny ticks. Talk about a buzzkill. Did I continue picking? Yes. Did I hurry up the process? Also, yes. This happened on the ridgeline by my house, so I hightailed it down the hill to make it home and scrub myself in the shower. So, with that being said, please resist the urge to roll around in a black trumpet patch in celebration because you just might pick up a bunch of hitchhikers you ain't wanting to give a ride home to.

Scan here to get a look at some black trumpets, plus a bonus stroganoff recipe!

Black trumpets range in color from dark black to ashy gray. Here, you can see their iconic trumpeted shape and rolled margins.

LOOKS:

Black trumpets typically grow to have a top width of 1 to 3 inches (2.5 to 7.5 cm), with varying sizes often found within the same patch. They are funnel-shaped and range from a smokey gray to dark black in color. Older specimens sometimes develop an orangey to yellowish coloring in different parts of the mushroom. The top portion has a circular shape and has a margin, or lip, that rolls downward. They are hollow throughout, mostly smooth in texture (lacking gills), but can also have somewhat wrinkly flesh. The skin is thin and brittle to the touch. The stem appears to be a continuation of the "cap" and tapers downward toward the ground.

STOMPING GROUNDS:

They often fruit amongst broadleaf trees and in mixed forests. Black trumpets also tend to love partying in the moss. The mnemonic device B.O.M.B. applies here, too, so look around beech, oak, and moss, baby.

GROWTH PATTERNS:

They often grow scattered or in colonies from the ground.

SPORE PRINT COLOR:

salmon to pinkish orange

FLAVOR NOTES:

earthy, rich, smoky

POTENTIAL LOOKALIKES:

The devil's urn mushroom (*Urnula craterium*) can possibly be mistaken for black trumpets. The devil's urn has a more cuplike shape and has thicker, firmer, rubbery flesh. Another key difference is that devil's urn grows in the spring while black trumpets will fruit in summer/fall. Devil's urn is not a toxic mushroom, but they definitely aren't as yummy as the black trumpet.

Blue chanterelles (*Polyozellus multiplex*) are another potential lookalike for black trumpets. Blue chanterelles have a more blueish hue than the black-to-gray black trumpet. Blue chanterelles are more wrinkled in appearance, whereas black trumpets' wrinkles are fainter. Blue chanterelles tend to grow in more tightly packed, dense clusters, and black trumpets are more scattered in their growth patterns. Blue chanterelles are also a choice edible mushroom.

RIGHT-QUICK RUNDOWN:

- trumpet- or bull-horn–shaped
- grayish black in color
- smooth (lacking gills)
- thin flesh
- hollow

Edible Mushrooms of the Summer 83

WHITE AND BLACK TRUMPET PIZZA

INGREDIENTS

Carmelized Onions
- 3 medium onions
- 2 tablespoons (28 ml) olive oil

Mushroom Mixture
- 8 ounces (225 g) clean black trumpet mushrooms
- ¼ teaspoon dried thyme
- ¼ teaspoon dried rosemary
- 1 tablespoon (14 g) butter

Sauce
- 1 tablespoon (14 g) butter
- 2 cloves minced garlic
- 6 ounces (170 g) softened cream cheese
- ½ cup (120 ml) whole milk
- ¼ teaspoon salt

Assembly
- 1 package (16 ounces, or 455 g) pizza dough (or 1 homemade dough)
- 1 cup (110 g) shredded fontina cheese

DIRECTIONS

1. Thinly slice the onions and sauté in a skillet on medium heat with olive oil for approximately 30 minutes, stirring constantly until golden brown and caramelized. Remove from skillet and set aside.
2. Add the mushrooms, thyme, rosemary, and butter to the same skillet you used previously and sauté on medium-high heat for approximately 10 minutes. Remove from the heat when finished.
3. To prepare the cream sauce, add 1 tablespoon (14 g) of butter and minced garlic to a small saucepot over medium heat. Sauté for 1 minute. Add the softened cream cheese, milk, and salt. Continue to heat and whisk constantly over medium heat until a smooth sauce forms, about 5 minutes. Remove from the heat and set aside.
4. Adjust the rack in the oven to the highest position and preheat the oven to 450°F (230°C, or gas mark 8).
5. Roll out the pizza dough and spread the cream sauce evenly on the dough.
6. Sprinkle half of the fontina cheese on top of the sauce, top with the mushroom mixture and caramelized onions, and sprinkle the remaining cheese on top.
7. Bake at 450°F (230°C, or gas mark 8) for 10 to 15 minutes, monitoring regularly to make sure it doesn't burn.
8. Remove from the oven, let cool, slice it up, and enjoy!

CHICKEN OF THE WOODS (GENUS *LAETIPORUS*)

Just like the previously discussed chanterelles, there are more than one type of chicken of the woods. In this text, we will cover the two most common and sought after chicken of the woods in my neck of the woods, *Laetiporus sulphureus* and *Laetiporus cinnicinatus*, although there are many other types that occur across the globe. This is a good reminder as to why Latin bionimals are sooo important! When we say the common name "chicken of the woods," we need to also clarify its Latin binonmial to know exactly which type we are referring to in order to avoid any confusion or any bubble guts.

Something of major note when it comes to this mushroom is that some folks can hack it and some folks just cannot. Research has shown that anywhere from 20 to 50 percent of people have some sort of sensitivity or adverse reaction to chicken of the woods. These reactions may include swelling, dizziness, headache, nausea, vomiting, and diarrhea. Like I confessed in earlier chapters, I am in that 20 to 50 percent of people when it comes to *Laetiporus sulphureus*, unfortunately. I've tried it a handful of times thinkin' maybe it was just the particular specimen, but nope, it ain't. I eat *Laetiporus cincinnatus* with no problems, however.

Something to take into consideration when preparing and trying this mushroom: You must make sure you cook it thoroughly. Undercooked chicken of the woods will likely cause GI issues. You should also eat only young, healthy specimens. Many folks opt to eat only the tender outer edges of the mushroom and avoid the inner, older parts of the fungus. As a general rule of thumb when I am harvesting, I like to utilize smaller, nubby mushrooms with a more supple texture. Ya don't want it to snap apart easily and kick up dust. I will use older chicken of the woods to dehydrate for seasoning, however. When I say older specimens, I do not mean rotten, bug-infested mushrooms, but rather I mean healthy, just a little larger and mature, specimens.

If you plan on preparing a meal including chicken of the woods for your dinner guests, you may want to reconsider, especially if these guests have never sampled these wild mushrooms. Y'all could end up with an unplanned game of digestive roulette if ya don't.

Now, I'm not tryin' to scare ya off from chicken of the woods, but am offering up my best, transparent advice so you can avoid what happened to me and many others who maybe didn't start *small* with a new-to-them wild mushroom. I have many family and friends who eat chicken of the woods no problem. I still forage and prepare it for those people even though I can't partake in the chow-down. It's still a huge rush when I find a big ol' flush of chicken of the woods regardless, so I recommend you give it a whirl—but do so cautiously.

Edible Mushrooms of the Summer

CHICKEN OF THE WOODS, YELLOW PORED
(*LAETIPORUS SULPHUREUS*)

The most common chicken of the woods I come across is this guy, *Laetiporus sulphureus*. Some folks call it "the yellow chicken of the woods." Chicken of the woods is another good beginner mushroom because nothing else looks like it. With its flashy bright oranges and yellows, it's easy to spot and hard to mistake. It shows up in massive overlapping shelves growing directly from hardwood, and I find it most frequently on oak trees. I remember the first time I found some chicken of the woods. I knew absolutely nothin' about mushrooms and thought to myself, "Woah! This *has* to be a poisonous mushroom!" I figured this due to the super vibrant coloring, but boy was I wrong! As I mentioned earlier, there is no one way to determine whether a mushroom is edible. Some might think, like I did as a novice, that bright colors mean to stay away—not in this case. This flashy mushroom is highly sought after by foragers and wild food chefs. Some restaurants will pay out the wazoo for a flush of chicken of the woods.

Remember, younger specimens of this mushroom seem to cause less of an angry gut. If you can find it fruitin' in its infancy, this is choice. Young chicken of the woods can appear to be small bubbly lumps or nodes. Think of it as lil' chicken nuggets, and you can even treat them as such by breading and frying them at home.

Scan here to see a *massive* chicken of the woods I found!

CHICKEN OF THE WOODS ALL-PURPOSE SEASONING SALT

INGREDIENTS
- 36 grams (1.3 ounces) dehydrated chicken of the woods mushrooms
- 72 grams (2.5 ounces) coarse sea salt
- 1 tablespoon (1 g) dried parsley
- ½ teaspoon roasted garlic powder
- ½ teaspoon onion powder
- ½ teaspoon pepper
- ½ teaspoon dried sage
- ½ tablespoon dried thyme
- 1 heapin' tablespoon (24 g) truffle zest

DIRECTIONS
1. Finely grind the dried chicken of the woods mushroom in a spice grinder until it is a fine powder.
2. In a bowl, mix the mushroom powder and all the other ingredients.
3. Grind it all together until fine.
4. Store in an airtight container up to six months. This makes about 1 cup (235 ml) of seasoning.
5. Add it to meat, pasta, eggs, or any savory dish, really, for an earthy, umami punch.

Go Forth and Forage

I consider this to be primo eatin' chicken of the woods. It's young, tender, and less likely to cause raw butt.

LOOKS:

Chicken of the woods typically grow to have fan-shaped, overlapping caps up to a foot (30.5 cm) wide. The cap is bright orange with yellow margins when younger, and the margins become wavy. The underside has teeny tiny pores (almost too small to see with the naked eye in a baby specimen) and is pale to bright yellow in color, lacking a stem (it's connected directly to the wood). When younger, the flesh is softer and has more give to it, and then it becomes harder, chalkier, and more stringy with age.

STOMPING GROUNDS:

They are found growing on both living and dead/dying hardwood trees and stumps, especially oak and beech.

GROWTH PATTERNS:

They can be found growing in overlapping clusters, sometimes singularly or scattered, from hardwood trees.

SPORE PRINT COLOR:

white

FLAVOR NOTES:

mild, slightly sour with lemony notes, meaty texture

POTENTIAL LOOKALIKES:

Laetiporus huroniensis is a nearly identical type of chicken of the woods, with the difference being the type of tree it grows from. *Laetiporus huroniensis* grows from conifer trees instead of hardwood trees, especially on hemlocks. See why knowing your trees is important? Edibility of this mushroom is up for debate. Some report they have eaten it without issue, but others state it should be avoided due to it containing toxins from the host tree. When edibility is questionable, I skip it! Given I have a sensitivity to *Laetiporus sulphureus*, ya definitely ain't gonna see me munchin' on any *Laetiporus huroniensis*.

Laetiporus cincinnatus is also a potential lookalike, but lucky for you, it's delicious and we are about to cover it in the next profile, so you'll know the difference!

RIGHT-QUICK RUNDOWN:

- bright-orange to yellow cap
- yellow underside with tiny pores
- fan-shaped
- no visible stem
- growing from wood
- growing in overlapping clusters

CHICKEN OF THE WOODS, WHITE PORED
(LAETIPORUS CINCINNATUS)

Now this here is *my* favorite chicken of the woods, *Laetiporus cincinnatus*, or as some people call it, "the white chicken of the woods." Not only is it easier on the guts, but I also think it might just be a bit purdier than *Laetiporus sulphureus*. This mushroom differs in that it has a peachier-colored cap and a porous white underside as compared to the bright orange top and yellow underside of *Laetiporus sulphureus*. This mushroom also showcases a different growth pattern. You'll often find them in gorgeous rosettes, with overlapping caps winding around, resembling a blooming rose. The white chicken of the woods is commonly found fruiting at the base of hardwood trees versus directly from the bark of the tree. You might sometimes find it sprouting up some distance from the tree, but it's actually growin' from buried wood or tree roots. If I haven't already sold you on leaning more toward this type of chicken of the woods, another bonus is that this type seems to have more of a resistance to the creepy crawlers. When I find the yellow chicken of the woods, it often comes with some beetles and bug holes, but with the white chicken of the woods, I notice they are much more pristine and virtually bugless. I find the texture of white chicken of the woods to be more tender and desirable, not turnin' as chalky and woody as the yellow chicken of the woods.

Note the unmistakable rosette shape and white margins of *Laetiporus cincinnatus* as key ways to distinguish it from *Laetiporus sulphureus*.

LOOKS:

The entire rosette structure can typically grow to be up to 25 inches (63.5 cm) across, with the individual caps ranging from 1 to 10 inches (2.5 to 25.5 cm) wide. The caps are zoned with pinkish orange to peach and white colors with the underside being white to cream-colored with very small pores. The caps have distinct white, wavy margins when younger, and then the margins become the same color as the cap when older. All fan-shaped caps are connected to a central stalk.

STOMPING GROUNDS:

They prefer to grow at the base or from the roots/buried wood near hardwood trees, especially oak.

GROWTH PATTERNS:

They can be found growing most commonly in a rosette shape, with overlapping levels of caps, typically as standalone structure, or as a few scattered structures if you're blessed enough.

SPORE PRINT COLOR:

white

FLAVOR NOTES:

mild, slightly sour with lemony notes (I think less sour than *Laetiporus sulphureus*), meaty texture

POTENTIAL LOOKALIKES:

When *Laetiporus cincinnatus* is older and faded to a lighter whitish color, it can resemble *Bondarzewia berkeleyi*, or the Berkeley's polypore mushroom. Both mushrooms grow in rosettes at the base of trees. However, Berkeley's polypore has larger, more distinguishable pores on the underside and has lighter brownish and cream zonations on the cap. I also tend to find multiple fruitings of Berkeley's polypore together, whereas I usually find white chicken of the woods as one single fruit. Berkeley's polypore is also edible, and we are gonna talk about them later, too!

You can see how one might think they've found an albino Laetiporus sulphureus *when it comes to this Berkeley's polypore, but white chicken of the woods does not fade to this degree of fairness, retaining its nice hues of pink and peachy orange, and will not have the pale brown and white zonations of the Berkeley's polypore.*

RIGHT-QUICK RUNDOWN:

- pinkish-orange to peach cap
- white underside with tiny pores
- fan-shaped
- central stalk present
- growing in a rosette with overlapping caps

Edible Mushrooms of the Summer

TAWNY MILK CAP (*LACTIFLUUS VOLEMUS*)

Here's a very underrated mushroom in my opinion—the tawny milk cap. It has a bushel of common names, including weeping milk cap, leather back, the orange-brown milky, and Bradley milk cap. I, however, have gifted it my own common name and that is—the crotch rocket. Why you ask? Well, when you cut this sucker, it smells just. Like. Fish. You do the math. This fishy odor is a certain feature in the identification of this milk cap. The fishy smell goes away with cooking, and it tastes nothing like fish, just a nutty, earthy mushroom. Now, when you collect ya a sack full of these crotch rockets, ya might be skippin' down the trail and pass some folks who are gonna do a double take and think you need to take a shower. But really, it's just your mushrooms. Pay those people no mind and who cares anyway? Ya got a yummy bounty in tow! Some people toss the stems of these mushrooms, claiming they are somewhat grainy in texture and unpleasant, but I cook and eat this mushroom from cap to stem and enjoy all parts.

Let's make "crotch rockets" happen. I've always wanted to name a mushroom.

Peep that milk. Very healthy specimens of crotch rockets exude a whole lotta mushroom milk and also bruise brown when damaged, as is present on the gills of the feller all the way to the left here.

LOOKS:

Tawny milk caps typically grow to be up to 5 inches (13 cm) wide and 5 inches (13 cm) tall. The cap is pale orange when younger, turning a darker orangey brown when more mature. The cap has a central depression that is often slightly darker in color. The gills produce latex when damaged and eventually stain brown after being cut. The latex, or "milk," is sticky and has a fishy odor. The gills are off-white in color, crowded, attached to the stem, and often forked. The stem is smooth and usually a lighter creamish to yellow-orange color than the cap. The inside of the stem may appear granular in texture.

STOMPING GROUNDS:

Tawny milk caps prefer to grow in coniferous forests, especially among pine and spruce trees, but they can also grow around beech and oak trees.

GROWTH PATTERNS:

They can be found growing singularly or scattered from the forest floor.

SPORE PRINT COLOR:

white

FLAVOR NOTES:

mild, nutty

POTENTIAL LOOKALIKES:

Lactarius corrugis is similar to the crotch rocket, but it has a darker, redder cap, is typically larger in size, and will usually have a cap that is more wrinkly. *Lactarius corrugis* is also edible!

RIGHT-QUICK RUNDOWN:

- pale-orange to dark-orange cap
- smooth stem colored slightly lighter than the cap
- off-white gills that exude milk when cut
- gills staining brown when cut
- fishy odor

Edible Mushrooms of the Summer

PUFFBALLS

To make sure your puffball is edible, you want a solid white interior when sliced in half, like this common puffball. If it has any brown, green, or yellow spots inside, it is past its prime.

Puffball mushrooms are super cool. There are a few different ones, and they can come in many different shapes, shades, and sizes. There's a general rule to determine whether a puffball is edible, and it applies across all the different puffballs we are about to cover in the next few profiles. As we know, depending on the maturation of the mushroom, it can be past its prime and not so great for eating. With puffballs, the way to tell it's down for the dinner plate is by cuttin' it in half. If the interior of the puffball is nice, clean, and white, it's ready to go. If the puffball has started to turn brown, yellow, or greenish inside, it's outside the yum yum window and you're gonna wanna pass on it. Not to worry though! There's still something fun to do with it, and it's something I like to do it with any old puffballs I find—spread the spores! If it's a big feller that's too old for eating, I punt it! It's amusing and productive to scatter mushroom love all over the place. If it's a smaller dude and not puntin' size, just give it a crumble and throw it into the air like fungi confetti. It makes any foray occasion a party!

The taste of a puffball is most likely not gonna blow your socks off. It's very mild and airy, so I call it "tofu of the woods." This blank slate makes puffballs a perfect canvas for cookin'. The puffball does need a little help with seasoning or sauce to give it some more oomph. I personally love the texture. It can stand up to roasting, sautéing, a hard sear, or even a good grillin'. I've also dehydrated puffballs before to create a puffball powder. I then use this powder to add to soups and stews for an umami flare. The puffball possibilities are endless!

GIANT PUFFBALL
(CALVATIA GIGANTEA)

Holy puffball, buddy! This guy is *huge*. The giant puffball is always a treat to find because just look at it! It seems unreal and looks like someone abandoned a big ol' volleyball on the ground, but it's a mushroom. I heard tell that one of the biggest puffers ever found was pert near 8 feet (2.4 m) in diameter and weighed in at about 48 pounds (22 kg). That's a whole lotta mushroom, honey. Now, I ain't never found one *that* big, but I have found me some whoppers in my day.

When preparing the giant form of the puffball, I do like to peel off the outer layer of skin since I find it to be a little rubbery for my palate. It's easy to do. I typically cut giant puffballs into slices (and admire the squeak that it makes when I wiggle my knife through). You can then just grab onto the outer skin and peel it away like an orange. You're left with lovely, nekkid puffball slices ready for cookin'!

Scan here to watch me forage and cook up some puffball steaks!

Edible Mushrooms of the Summer

Puffballs are an easy one to spot because they will pop up seemingly overnight often in open fields and yards.

LOOKS:

Giant puffballs typically grow to be anywhere from 6 to 20 inches (15 to 51 cm) wide/tall but can absolutely get larger. They will be rounded, smooth, and white to off-white in color. In older puffers, you might find cracks or dimples in them. The stem is essentially nonexistent, and it will be attached directly to the ground. When cut in half, the inside flesh is white when fresh and young, developing yellowish to green spots when more mature.

STOMPING GROUNDS:

They prefer to grow in open areas like meadows, pastures, lawns, and at the opening of forests around hardwood trees.

GROWTH PATTERNS:

They can be found growing singularly or scattered in small groups.

SPORE PRINT COLOR:

greenish brown

FLAVOR NOTES:

not much flavor, but *slightly* earthy

POTENTIAL LOOKALIKES:

Since giant puffballs get, well . . . giant, there's not a lookalike to be particularly concerned about since it gets so large like no other rounded white mushroom does. In a giant puffball's infancy as a teeny baby, a potential lookalike could be earthballs (*Scleroderma citrinum*) or baby *Amanita* mushrooms. The easy way to tell the difference is earthballs will be black inside when sliced in half, while the puffball will be white, and baby *Amanitas* will have the outline of a mushroom inside when sliced, while the puffball will not. Earthballs and small *Amanitas* will not get as large as a giant puffball. Both the earthballs and baby *Amanitas* should not be consumed!

RIGHT-QUICK RUNDOWN:

- round, smooth, whitish in color
- large
- white, spongey interior
- lacking stem

GRILLED PUFFBALL STEAKS

INGREDIENTS
- 1 giant puffball mushroom
- Olive oil
- Garlic powder, salt, and pepper to taste*

DIRECTIONS
1. Cut large slices approximately 1-inch (2.5 cm) thick of puffball mushroom (these will become thinner when cooking) and peel off the outer layer.
2. Drizzle both sides of the puffball steaks with olive oil and season with garlic powder, salt, and pepper.
3. Place the puffball steaks on a grill preheated to 400°F (200°C).
4. Cook about 2 to 3 minutes per side, pressing down with a spatula and allowing each side to get some nice grill marks.
5. Serve and eat immediately.

*You can season the puffball steaks with whatever spices you like!

I scored a happy lil' puffball family, all beautifully pristine inside. You can see the range in size. I turned some into grilled steaks, some into puffball powder, and some into seared cubes in stir-fry.

Edible Mushrooms of the Summer

COMMON PUFFBALL
(*LYCOPERDON PERLATUM*)

Here's another puffer for ya, the common puffball. Guess what? It's common. I find these all over the dang place when I'm hikin' in the hills. You're liable to find them scattered all over the trailside. These are much smaller than the daddy giant puffball, usually being about the size of a quarter or up to golf ball–size. These lil' guys are cool because they look as if they are wearin' a little suit of armor with their spiky, studded exterior. Another common name for them is the *gem-studded puffball*. I'm tellin' ya, these common names make a lot of sense. Here's another one for you. The Latin name *Lycoperdon* translates like this: "lyco" means wolf and "perdon" means to fart. So, puffballs are wolf farts. Also, an appropriate name since when you find an older, mature puffball, it will often have a little hole or opening on top. This is how they spread their spores. You can give the puffball a squeeze to release spores into the air, or to make it fart. I love science.

When you're pickin' a mess of these, since they are on the smaller side, you'll have to gather a quite a few to make a dish worthwhile. These don't require peeling like the giant guy, despite their blingy coats. You'll also probably notice that when you're getting a handful of these, you'll get some sprinkley residue in your paws. It's totally normal! Forage on.

You can see here just how dainty and varying in size the common puffball can get. These were young, small specimens, so I had to do a lil' pickin' to get enough to cook up.

LOOKS:

They typically grow 3 inches (7.5 cm) tall/wide. The fruiting body is shaped like an upside-down pear. It is white to off-white in color and has tiny studs or spikes on the outer later, becoming less apparent in older specimens. When sliced in half, the inside is spongy and white, becoming more powdery and greenish brown with age.

STOMPING GROUNDS:

They prefer to grow on dead trees, in grassy clearings, and along the sides of trails.

GROWTH PATTERNS:

They can be found growing singularly, scattered, or gregariously.

SPORE PRINT COLOR:

olive brown

FLAVOR NOTES:

not much flavor, but *slightly* earthy

POTENTIAL LOOKALIKES:

Scleroderma citrinum, or the earthball, is a lookalike of the common puffball. The earthball is a toxic mushroom, but it's very easy to tell the difference in these when you slice it open. An earthball will be brown to black on the inside, whereas the puffball should be white. The earthball mushroom's exterior is also more brownish and speckled. Earthballs are much firmer to the touch than the spongier, squishy puffball.

Another possible lookalike of the common puffball can be baby *Amanita* mushrooms. *Amanitas* contain toxic mushrooms, so ya gotta be careful. The death cap mushroom, *Amanita phalloides*, starts out as a little egg that can resemble a small common puffball. To tell you have a puffball versus a deadly *Amanita*, slice it in half. If you see the outline of a baby mushroom inside, it's likely an *Amanita*. If it is solid white mass throughout, it's a puffball.

As seen here, the interior of earthball mushroom is dark brown to black in color. **Don't eat it!*

RIGHT-QUICK RUNDOWN:

- white, studded exterior
- white, spongy interior
- rounded with slightly tapering lower portion (pear-shaped)
- quarter- to golf ball–size

BRAIN PUFFBALL
(CALVATIA CRANIIFORMIS)

Appropriately named both commonly and scientifically, these puffballs look like, you guessed it, a daggone brain! Another common name is the *skull-shaped puffball*. It really *is* skull-ish shaped. It has a rounded top portion with a thinner portion below that. Think of an upside-down pear. It carries all those iconic puffball traits of being round, sprouting from the ground, and is white to tannish in color with the white interior. It's smaller than the giant puffball and has a wrinklier surface, resembling the lobes of a noggin noodle. It's totally freaky looking and totally edible.

LOOKS:

They typically grow to be 3 to 7 inches (7.5 to 18 cm) wide and the biggest upper portion of the mushroom is rounded. The exterior is white to tan in color, smooth when young, and then eventually develops wrinkles and cracks with age. The wrinkles resemble lobes of the brain. It is skull-shaped or shaped like an upside-down pear. The interior is white and spongey when prime, later turning green and yellow when past prime.

STOMPING GROUNDS:

They prefer to grow in open fields and meadows and in open hardwood areas of the forest.

GROWTH PATTERNS:

They can be found growing singularly, scattered, or gregariously.

SPORE PRINT COLOR:

yellowish brown to olive brown

FLAVOR NOTES:

not much flavor, but *slightly* earthy

POTENTIAL LOOKALIKES:

Just as with other puffballs listed in this book, earthballs (*Scleroderma citrinum*) and baby *Amanitas* can bear some resemblance. Cut the puffball in half to ensure it is solid white throughout and not a toxic lookalike!

RIGHT-QUICK RUNDOWN:

- white to tan in color
- upside down pear shape
- cracked, wrinkly/lobey exterior (like a brain)
- white, spongy interior

PEAR-SHAPED PUFFBALL
(APIOPERDON PYRIFORME)

These puffers are similar in size to common puffballs but have just a few differences. You'll find these balls oftentimes hanging out directly on fallen wood versus the other puffers that fruit from the ground. Pear-shaped puffballs can often be found partyin' with a lot of their buddies in thick, tightly packed clusters.

LOOKS:

They typically grow to be 1 to 3 inches (2.5 to 7.5 cm) wide and have a white exterior, becoming tan to brown with age. They are rounded/pear-shaped and the exterior may appear smooth or cracked depending on maturation. The interior is white when prime, turning green/yellow when past prime.

STOMPING GROUNDS:

They prefer to grow on both fallen hardwood and conifer wood. They also can be found growing in mulch.

GROWTH PATTERNS:

They can be found growing gregariously or in tight clusters.

SPORE PRINT COLOR:

olive brown

FLAVOR NOTES:

not much flavor, but *slightly* earthy

POTENTIAL LOOKALIKES:

See brain puffball profile at left.

RIGHT-QUICK RUNDOWN:

- whitish to tan in color
- rounded to pear-shaped
- growing on wood
- smooth, cracked, or granular texture on exterior (dependent on age)
- white, spongy interior

Edible Mushrooms of the Summer

BEEFSTEAK POLYPORE (*FISTULINA AMERICANA*)

This beefsteak polypore is showing off its red, gelatinous cap. Note the tiny bumps and how the cap resembles a tongue!

Now, this is one cool and funky mushroom. You may have guessed from its common name, *beefsteak*, that it just might have some meaty qualities—and you'd be right! This red hunk of fungus has numerous unique features, making it another easy beginner's mushroom! It also occurs fairly commonly east of the Rocky Mountains.

It has a look to it that is very similar to a slab of raw beef. When you cut into it, you see white streaks, or marbling, and it oozes a red liquid (think blood) when you give it a squeeze. Is that common name making sense now?

This is one of those wild mushrooms that breaks the rules in that it *can* be consumed raw. It can be added in its raw, fresh state to salads for a punchy, tangy flavor. Many folks like to marinate this mushroom overnight to take on different flavor profiles if they are planning on doing a heat preparation with it. It makes for a dang good meat substitute for obvious reasons, so whip up a tasty vegan dish with this feller!

Due to this mushroom's acidic flavor, it is best mixed with other wild mushrooms, so the citric flavor is not overpowering in your dish. Some folks love the interesting flavor, some aren't huge fans, but it's definitely worth a shot to see what ya think!

Beefsteak polypores have a light-colored, porous underside made up of teeny tiny holes.

LOOKS:

They typically grow to be up to 10 inches (25.5 cm) wide (sometimes larger) in a semi-circular fan shape, flattening out more as they mature. When young, the top has a bright red, gelatinous, sticky top and a bumpy texture like tiny taste buds that is very reminiscent of a beef tongue. They are awful purdy and glistening after a summer's rain. It has a fleshy appearance you can wiggle, jiggle, and move around when you touch the cap. The color of the cap becomes darker with age. The underside is lighter in color than the top, usually cream colored to off-white to yellowish. The spore-producing surface is porous and will actually expose tubes that bruise reddish brown when cut into a cross section. Sometimes a stalk is present, but more frequently the stalk is absent. If present, the stalk is dark, lateral to the cap, and short and stubby.

STOMPING GROUNDS:

They often grow at the base of living broadleaf trees and on stumps, especially oak trees.

GROWTH PATTERNS:

It can grow singularly or in small overlapping groups directly from wood. They can grow from summer to fall, but are most abundant in the summer.

SPORE PRINT COLOR:

pinkish salmon to pinkish brown

FLAVOR NOTES:

acidic, tart, citric, slightly fruity

POTENTIAL LOOKALIKES:

Some reishi mushrooms (*Ganoderma* sp.) could appear beefsteaklike from a distance since reishi can have bright red colors, can be fan-shaped, and can grow from wood. Upon inspection, you will quickly find out reishi are much harder in texture and they do not possess those meaty attributes of the beefsteak. So, fear not! You likely won't confuse this mushroom with other species due to its numerous distinctive features.

RIGHT-QUICK RUNDOWN:

- red, gelatinous/sticky cap
- white to off-white porous underside
- stem often absent
- growing from wood

OLD MAN OF THE WOODS
(STREOBILOMYCES FLOCCOPUS)

Okay, okay. This mushroom is not bringin' home any culinary awards of any kind anytime soon, but would you just look at old man of the woods?! He is one of my absolute favorite mushrooms to stare at, so I felt he deserved an honorable mention.

He is a bolete, so you'll find his underside is very porous and dark, kinda like an ol' dirty sponge. Ironically, that's also kinda what he tastes like. But, oh, that cap! How neat is it? It's a spiky little hat he wears that sets him apart from all the other boletes. Every time I find one, I find myself singing Neil Young.

Even if I don't recommend bringing this feller home to eat, I can say, he's super easy to identify and is a good beginner's mushroom—*and* if you're ever stranded and starving in the woods somewhere, just know you could roast this guy up as a last resort.

Here is a real old man of the woods. You can tell he's just about give out with his sheddin' cap and larger pores.

LOOKS:

They typically grow to have a cap width of up to 6 inches (15 cm) wide and they stand up to 6 inches (15 cm) tall. The cap is smokey gray as the base color with black woolly, spikey scales that can be pointy or more flattish. The underside reveals a gray to blackish spore-producing surface with many tiny pores. The stem is solid, uniform in shape, and colored similarly to the cap with a shaggy appearance. Sometimes, there is a ring of sorts visible on the stem as a hairy, gray annulus. When doing a cross section, the flesh stains orangish red.

STOMPING GROUNDS:

They prefer to grow in hardwood forests, especially around oak trees.

GROWTH PATTERNS:

They can be found growing solitarily or scattered from the forest floor.

SPORE PRINT COLOR:

dark brown to black

FLAVOR NOTES:

old, dirty dish sponge

POTENTIAL LOOKALIKES:

Ain't none!

RIGHT-QUICK RUNDOWN:

- grayish cap with spikey black scales
- dark, porous underside
- gray to black shaggy stem
- stains orangish red when doing a cross section

SHAGGY MANE (*COPRINUS COMATUS*)

Shaggy manes are a wild lookin' lil' mushroom, serving member-of-British-parliament-wig vibes. They are a bit of a fickle mushroom with a short lifespan. They are part of the ink cap mushroom family, meaning they start to turn ooey-gooey inky and dissolve over time. The texture ain't for everyone. When ya pick ya a mess of these, you want to get home ASAP to get cookin'. They can become squishy, watery, and unpleasant pretty quickly. Picking younger, firmer specimens that show no sign of inky-ness is the way to go to avoid any troublesome mush in the kitchen. Don't let that throw you, though. Shaggy manes can still be prepared into a delicious dish!

This here's a shaggy mane looking the way they should look when you wanna harvest them. You wanna make sure to pick the shaggy manes that are younger and firmer in texture, and ones that have not started to turn inky yet.

LOOKS:

They typically grow to be up to 8 inches (20 cm) tall and around 3 inches (7.5 cm) across. They have a very long, cylindrical cap that can start out covering the entire stem of the mushroom. The cap develops white scales that resemble wisps of hair. The margin of the mushroom starts to become black, inky, and will liquify, dissolving up the cap over time, exposing the stem more as it dissolves. The spore-producing surface is gilled and crowded. The gills start out white, turn pink, then finally become black when inky. The stem is hollow inside with some fibrous strands within.

STOMPING GROUNDS:

They prefer to grow on roadsides, along the sides of trails, in lawns, in woodchips, on disturbed ground, and even in gravel.

GROWTH PATTERNS:

They can grow in clusters, scattered, or gregariously.

SPORE PRINT COLOR:

black

FLAVOR NOTES:

mild, earthy

POTENTIAL LOOKALIKES:

Coprinopsis atramentaria, or the common ink cap, may resemble a shaggy mane. The common ink cap has a smoother cap surface, lacking the "shagginess" or scales of the shaggy mane. Common ink caps are usually smaller in size and grow more often in tighter clusters than shaggy manes. Eating common inkcaps is not recommended.

RIGHT-QUICK RUNDOWN:

- white, wispy cylindrical cap
- inky margin (when older)
- white to pink to black gills (depending on age)
- long, hollow stem with some fibers inside

REISHI MUSHROOMS (GENUS *GANODERMA*)

Just like some of the previously discussed mushrooms, what we refer to as reishi mushrooms have a few different types within their happy family. *Ganoderma* mushrooms share some characteristics, while also having differences to separate the different kinds. For the most part, reishi mushrooms have the following characteristics in common: growing in conklike brackets from wood; a tough, fan-shaped, lacquered/shiny cap; and cap colors ranging from deep maroon to red to orange to a whiter outer cap margin, sometimes with zonations. Reishis are also a type of polypore mushroom, with tiny pores on the underside of the cap.

I use my reishi mushrooms medicinally, making teas and tinctures to harness their lovely healin' powers. Reishi mushroom benefits may include boosting immunity, improving energy levels, reducing anxiety/stress, promoting heart health, reducing inflammation, stabilizing blood sugar levels, and so much more! As always, be sure to check with your doctor before trying any new mushroom supplements to ensure they are safe for you and will not interact with any medical conditions you may have or any prescribed medications you may be taking.

Ganoderma aren't really a culinary mushroom. It ain't really anything you wanna go chewing on or you'll be chewing until pigs start flyin'—or until you dislocate your jaw, whichever comes first. Plus, it's purdy bitter on the ol' palate.

With all that being said, you always want to confirm your reishi ID, as is the case with any mushroom. I have found a few different types of reishi around my homebase, but I will be covering the one I find most often in the following profile.

HEMLOCK REISHI (*GANODERMA TSUGAE*)

As I yacked about previously, there are a few different types of reishi, but this particular one is what I find most often. A lot of my favorite hikin' trails happen to be amongst lots of hemlock trees, and *Ganoderma tsugae* are fruits from the hemlocks! This is probably why I find it most often, so keep your eyes peeled for other reishi when you're out and about because they are definitely out there.

Here, we have two different ages of reishi. The younger ones can sometimes look like a small white blob when they first start growing. Notice them shiny zonations!

LOOKS:

Hemlock reishi typically grow to be up to 1 foot (30.5 cm) wide and the cap is often fan-shaped with shiny, varnishy colors on the cap, ranging from red or brown to orange yellow and often with a white margin at the cap. The cap colors become darker with age. The underside has teeny tiny pores, is white to off-white, bruises when damaged or scraped, and turns darker with age. The stem is shiny, growing up to 7 inches (18 cm) in length, and depending on how it is growing from the wood, it can be centrally arranged to the cap (when growing straight up and down) or be offset to the side of the cap (when growing sideways out of a tree). The mushroom's flesh is firm, but with a little give when younger, and it becomes very hard and tough when older.

STOMPING GROUNDS:

They fruit directly from wood of especially hemlock trees and sometimes other conifers; very seldom on hardwood trees.

GROWTH PATTERNS:

They can grow solitarily, gregariously, or sometimes in stacked clusters from wood.

SPORE PRINT COLOR:

brown

FLAVOR NOTES:

bitter

POTENTIAL LOOKALIKES:

As I mentioned previously, there's a lot of different types of reishi, and they all carry similar characteristics. Common lookalikes of hemlock reishi would be other reishi including *Ganoderma curtisii*, *Ganoderma sessile*, and *Ganoderma lucidum*. *Ganoderma curtisii* is usually smaller in size and grows on hardwoods and *Ganoderma sessile* and *Ganoderma lucidum* also like to grow from hardwoods. Since hemlock reishi grows on hemlock trees, it is easy to differentiate. Fear not, all the reishi mentioned here can be used medicinally.

RIGHT-QUICK RUNDOWN:

- shiny cap with colors ranging from maroon, red, orange, and brown with a white cap margin (when young)
- growing from hemlock/conifer tree
- white to off-white porous underside
- corky to hard texture

REISHI TEA

INGREDIENTS
- Dried reishi mushrooms (see instructions at right)
- Hot water
- Sweetener/spices of choice (optional)

DIRECTIONS

To Dry the Reishi Mushrooms:

1. To the best of your ability, carefully cut up the reishi mushrooms into small pieces. If you're not able to cut them very small, you can dry them whole and attempt to grind or blend later. Just know they are tough lil' buggers and require a quality grinder/blender.
2. Dry on a drying rack or in a dehydrator until no moisture remains. The length of time will vary depending on the size/thickness of the pieces.

To Make the Reishi Tea:

1. To a small pot, add filtered water and 1 to 2 teaspoons (1 to 2 g) of dried reishi mushrooms per 1 cup (235 ml) of water.
2. Bring the water and reishi to a boil and then lower the heat to simmer for at least 20 minutes.
3. Strain and pour into a cup.
4. Add sweeteners and spices of your choice, as reishi tea can taste bitter.
5. Sip and enjoy!

Note: You can use fresh reishi if desired, but I like to dry it to have it at the ready whenever I want a cup. If using fresh reishi, chop or slice the mushrooms, add to a pot of filtered water, bring to a boil, and simmer for 20 minutes to up to 2 hours, depending on how strong you like the brew, before straining.

Edible Mushrooms of the Summer

BERKELEY'S POLYPORE
(BONDARZEWIA BERKELEYI)

No, this ain't an old chicken of the woods, this is Berkeley's polypore! Some people might see it from a distance and think it's a *Laetiporus cincinnatus* that has baked in the sun too long, and that makes sense given its growing pattern. It likes to grow in rosettes at the base of trees, but it is its own edible mushroom. It is extremely common come early springtime but is most abundant in the summer. Then, they make their last appearances in the early fall. If you find one, mark your spot and your calendar because it's likely to fruit in the exact same spot near the same time the following year.

This is one you'll want to eat when fresh and young. Like chicken of the woods, Berkeley's polypore can become one tough and bitter Betty with age. Gettin' the timing right on this one can be tricky. If it's huge, it's likely too chewy and bitter. Harvest the tender outer edges that your knife easily glides through and leave behind the tougher, woodier parts.

BERKELEY JERKY

INGREDIENTS
- Approximately 4 cups (300 g) Berkeley's polypore caps, cut into strips

Marinade:
- 2 tablespoons (28 ml) olive oil
- 2 tablespoons (40 g) honey
- 2 teaspoons (10 ml) low-sodium soy sauce
- 2 teaspoons (10 ml) liquid smoke
- ½ teaspoon apple cider vinegar
- ½ teaspoon garlic powder
- ½ teaspoon salt
- ¼ teaspoon paprika
- ½ teaspoon onion powder
- ½ teaspoon red chili flakes

DIRECTIONS
1. Bring a large pot of water to a boil. Add in the mushroom strips and boil for 10 minutes.
2. Strain the mushrooms, let cool, and dab dry with paper towels once cooled.
3. In a large bowl, prepare the marinade by whisking all the ingredients together until well combined.
4. Add the cooled mushroom strips to the marinade, mix well, cover, and put in the fridge for at least 24 hours to marinate.
5. After 24 hours, place the marinated mushroom strips into a single layer on dehydrator trays and dehydrate for 6 to 8 hours or until they are the texture ya like.
6. Bag 'em up and prepare to wow everyone who eats it and thinks it's meat jerky, but it's *mushrooms*!

Berkeley's polypore is within the polypore family since it sports an underside full o' pores and has a thicker, leathery texture to it.

LOOKS:

The whole mushroom mass typically grows to be up to 2 to 3 feet (61 to 91.5 cm) wide and grows in a rosette pattern with whitish cream to gray layering with fan-shaped overlapping cap clusters. The cap can have light zonations present when younger, and the cap color darkens with age. The underside is white to cream colored with many tiny pores. The short, central or slightly off-center stem is anchored into the ground and oftentimes not extremely visible when peeking at the mushroom from the side.

STOMPING GROUNDS:

They prefer to grow from the ground at the base of hardwood trees, especially oak trees.

GROWTH PATTERNS:

They can be found growing solitarily or gregariously at the base of hardwood trees/stumps.

SPORE PRINT COLOR:

white

FLAVOR NOTES:

mild, slightly bitter (more bitter when older), woody

POTENTIAL LOOKALIKES:

Meripilus sumstinei, or the black-staining polypore, can resemble Berekley's polypore. The main, easy way to tell the difference is all in the name. Black-staining polypore will bruise or "stain" black when cut, and Berkeley's polypore will not. The Berkeley's polypore also grows to be much larger than the black-staining polypore. Just like Berkeley's polypore, black-staining polypore is edible when young!

Note the darker edges on this black-staining polypore.

RIGHT-QUICK RUNDOWN:

- growing at base of hardwood trees/stumps
- large, layering rosette growth pattern
- caps are whitish to cream to tan in color
- white to cream porous underside
- does not turn black when cut

You can use the recipe for Berkeley Jerkey at left for other mushrooms, too. Some good candidates would be hen of the woods, oysters, chicken of the woods, black staining polypore, or basically any edible wild mushroom that gets real big.

INDIGO MILK CAP (*LACTARIUS INDIGO*)

This older indigo milk cap shows how the purple/blue zonations dull with age and the mushroom becomes more grayish silver with only hints of indigo.

Say howdy to one of my favorite mushrooms to find and also another mushroom that really got me into this whole thang that I do—the indigo milk cap! I remember when I started trying my hand at mushroom huntin', I had seen other people post pictures of this feller and I wanted to find one so badly. To me, it seemed like a unicorn because I had never noticed one before. I didn't think I'd ever find it, but I sure did and rather quickly to my surprise. It's actually purdy abundant in the summer.

Like most milk caps, these guys like to grow around conifer trees, especially pine, but they also take a likin' to oaks and beech trees. I often find mine pokin' out of leaf litter. Their vibrant, surreal color makes them a breeze to spot.

This mushroom has had some trash talked about its culinary value. Some consider them to be too grainy in texture to eat, but I find it to be just fine texturally and in flavor. As much as I would love for these mushrooms to retain their bright, indigo color when cooked, they just don't. It is unfortunate, but they turn more of a gray color after layin' some heat to 'em. Yes, they do have milky gills, but this latex goo gets all cooked up and is just fine, lending no weirdness to a dish. Indigos stand up well to a sauté or roasting and pair well with herbs and garlic (but what doesn't?). Once the mushroom starts to lose its bright indigo color, it develops a more bitter taste. Harvest young, healthy specimens for an optimum flavor experience.

The indigo milk cap is a perfect example of how mushrooms can come in so many different neat colors and look like something straight from *Alice in Wonderland*. Its purple/blue color makes it so whimsical to me and the bonus of knowing I can eat it don't hurt either! Most guidebooks refer to this mushroom as being blue in color, but to me, it's purple. Whatever color you see, I think we can agree it's downright gorgeous.

The indigo milk cap is sure hard to mistake with its indigo gills and indigo milk!

LOOKS:

They typically grow to be 2 to 6 inches (5 to 15 cm) across the cap with the stem being around 2 to 4 inches (5 to 10 cm) long. The cap has purple and lighter indigo/gray zonations when younger, eventually fading to be more grayish with age. The cap, especially when mature, usually has a depressed center and loves to catch rainwater. The gills are also indigo in color and will exude an indigo milk when cut/damaged, and the milk can stain greenish. The gills are crowded under the cap. The stem can be somewhat stubby in appearance and is often colored similarly to the cap, sometimes having little indigo speckles on it. When cut in half, the inner flesh of the mushroom is silvery-gray and can turn green when damaged, like the gills.

STOMPING GROUNDS:

They prefer to grow around both hardwoods and conifers, especially pine and oak trees, in mixed forests.

GROWTH PATTERNS:

They can be found growing solitarily, scattered, or gregariously from the ground, moss, or leaf litter.

SPORE PRINT COLOR:

cream

FLAVOR NOTES:

nutty, earthy, a lot like button mushrooms, but slightly grainier

POTENTIAL LOOKALIKES:

At a glance, wood blewit mushrooms (*Lepista nuda*) or some purple *Cortinarius* mushrooms could trick ya into thinkin' ya got an indigo milk cap. Tellin' the difference couldn't be easier. Cut the gills. If the mushroom is blue and produces milk, ya got an indigo milk cap. If it doesn't secrete milk, then it may not be your guy. Wood blewits and *Cortinarius* sp. also are more purple in color, whereas the indigo milk cap is more blueish.

RIGHT-QUICK RUNDOWN:

- indigo to grayish zonated cap
- depressed cap center
- crowded indigo gills
- exudes indigo milk, later staining greenish
- turning paler, gray/silver with age

Edible Mushrooms of the Summer

OYSTER MUSHROOMS (GENUS *PLEUROTUS*)

There are a few different types of oyster mushrooms out there. They are all the rage and rightly so. They are easy to ID, plentiful, meaty, and yummy. The genus *Pleurotus* encompasses the oyster mushroom family. *Pleurotus* means "side ear," which makes a lot of sense when you think about an oyster mushroom. Oyster mushrooms' stem or stalk (when obviously present) is off-center or set to the side of the cap of the mushroom. Oyster mushrooms also have those decurrent gills we talked about, or gills that run down the stem. They like to grow in overlapping clusters from hardwood trees, especially ones that are dead/dying. The size and cap color varies depending on what type of oyster mushroom it is. All true oyster mushrooms are edible, so if you mistake one true oyster for another type of true oyster, don't be scurred. You can eat it.

Oyster mushrooms have a mild, umami, slightly seafood taste to them. Many oysters also have a slight smell of anise, think licorice-ish, along with a slightly fishy smell. I know that sounds weird and unappetizing, but I promise they are good. Trust momma.

Another fun fact about oyster mushrooms is that they are one of the easiest fungi to grow at home! I love growin' gobs of oyster mushrooms to utilize and preserve for myself and my buddies.

Some oysters I find the most and that occur frequently in the eastern United States include: *Pleurotus ostreatus* (probably the most common), *Pleurotus pulmonarius*, *Pleurotus dryinus*, and *Pleurotus populinus*.

We will cover the most common and tasty oysters in this here guide. We shall start with the summer edition of the oyster in the profile to follow and later cover the more common oyster, *Pleurotus ostreatus*, in the fall chapter.

SUMMER OYSTER
(*PLEUROTUS PULMONARIUS*)

This type of oyster starts to fruit in the summer months into the early fall and differs from the more common *Pleurotus ostreatus* in that it is smaller in size, thinner, lighter white in color, and likes warmer temperatures. I've also noticed that summer oysters, at least in my experience, tend to have a more prominent stem and also have a more scattered growth pattern than the *Pleurotus ostreatus* I usually find growin' in giant overlapping clusters. The good thing is, they don't lack in flavor. Yeehaw!

Edible Mushrooms of the Summer

Note the summer oysters' more pearly white shades and more distinct off-center stem.

LOOKS:

Each single oyster typically grows to be 2 to 5 inches (5 to 13 cm) across and is lung-shaped initially, becoming more fan-shaped and flattening with age. Eventually, the margins can become wavy when very mature. They are white to off-white to tan in color. The underside has decurrent gills that are white to off-white when fresh, browning or yellowing with age. The stem, if present (which is usually more noticeable than with other oysters), is stubby and off-center, unless the mushrooms are fruiting from atop a log, then the stem can sometimes be more centrally located with the actual mushroom appearing to grow more circular.

STOMPING GROUNDS:

They prefer to grow from the wood of hardwood trees, especially oak and beech.

GROWTH PATTERNS:

They can be found growing in scattered clusters or overlapping shelves from hardwood trees.

SPORE PRINT COLOR:

white to cream to light purple

FLAVOR NOTES:

mild umami flavor, nutty, faint smell of anise

POTENTIAL LOOKALIKES:

There's a common lookalike for the oyster mushroom called *Plueocybella porrigens*, or the angel wing mushroom. The main, key differences are what kind of tree they are fruitin' from and what time of year you find them. Angel wings are commonly found growing from conifer trees in the colder months, whereas summer oysters are found on hardwood trees in the warmer months. Angel wings also tend to develop a more trumpetlike shape and are smaller, whiter, and thinner than your average summer oyster. Edibility is still debated today on angel wings. Some eat 'em with no issues, some say not to eat 'em, so you know what I'll say ... nope!

The angel wing mushroom has a more trumpeted appearance and sprouts from conifers in the wintertime.

RIGHT-QUICK RUNDOWN:

- white to off-white to tan fan-shaped caps growing in clusters/shelves
- decurrent gills
- growing from hardwood
- growing in summer months
- slight odor of anise

WILD SUMMER MUSHROOM TOAST

INGREDIENTS
- 2 tablespoons (28 g) butter
- 1 large shallot, thinly sliced
- 4 cloves minced garlic
- 4 cups (weight will vary) fresh foraged summer mushrooms, roughly chopped (You could use chanterelles, black trumpets, oysters, milk caps, chicken of the woods, etc.)
- 4 thick slices sourdough bread
- 2 large handfuls of spinach
- Salt and pepper to taste
- ½ cup (125 g) ricotta cheese (or more if ya wanna lay it on thick)
- ½ cup (56 g) shredded gouda

DIRECTIONS
1. In a large skillet on medium-high heat, melt the butter and add the shallots to sauté for 2 minutes.
2. Add the minced garlic and sauté for another minute.
3. Add the cleaned, chopped mushrooms. Sauté for 10 minutes.
4. While the mushrooms are cooking, add the sourdough slices to a preheated 350°F (180°C, or gas mark 4) oven and toast until they are done to your liking.
5. After 10 minutes of mushroom cookery, throw the spinach into the skillet and let it wilt. Add salt and pepper to taste.
6. When the bread is done, smear the toast with ricotta cheese and hit it with some additional salt and pepper.
7. Top the toast with your hot mushroom mixture. Don't be shy. Pile it high!
8. Sprinkle the gouda cheese on top. Place the dressed-up toasts back in the oven until the cheese is just melty. Enjoy!

ROOTED AGARIC (*HYMENOPELLIS FURFURACEA*)

Woah, buddy! Check out the legs on this'un! This is the rooted agaric mushroom. I'll admit, it's not the *best* tasting mushroom, but it's a super common one you'll find dotting the forest in the summer, and I just want you to know what it is and that it is in fact an edible species! Most folks skip eating the stem and only eat the more tender cap. The stem is purdy fibrous and does not make for a good mouthfeel, so I'd have to agree that the caps are really all I wanna put in my mouth.

How does it acquire the name rooted agaric, you ask? Well, if you dig down on this sucker, it will have a looong, tapered root sunk deep into the ground. I can't help but check out the root on these every now and then. It's like a fun little treasure hunt to dig down and find the end.

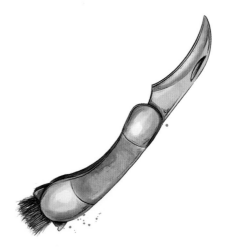

Get a load of that big ol' root!

Just so you can get a gander at that stem, here's a rooted agaric in my hand for scale. The cap will turn upward as you can see here once it becomes a more mature mushie.

LOOKS:

The cap typically grows to be up to 6 inches (15 cm) wide and is beige to tannish brown in color. The cap may start out bell-shaped when young, eventually flattening as it ages, and then turning upward when very mature. The cap has a bump in the center (an umbo) that is usually darker in color than the rest of the cap. The cap develops wrinkles or ripples around the central bump. Some consider this pattern to be almost snakeskinlike. The cap often appears to be shiny/wet when fresh. The underside reveals white to off-white gills attached to the stem. The stem can grow to be up to 1 foot (30.5 cm) long, is fibrous/stringy, and is usually uniform in shape/thickness. The upper portion of the stem tends to be lighter white/off-white in color, darkening to more of a tan/brown as you follow it down to the ground. The root below ground can vary in length, but usually it's purdy dang long when excavated completely!

STOMPING GROUNDS:

It can appear to be growin' directly from the ground, but it is actually growing from buried, dead wood. Other times, it grows directly off the dead wood.

GROWTH PATTERNS:

They can be found growing solitarily, or in small groups from the ground near rotting hardwoods, or directly from rotting wood/stumps/logs.

SPORE PRINT COLOR:

white

FLAVOR NOTES:

mild, slight seafood taste

POTENTIAL LOOKALIKES:

With legs like that, there ain't none!

RIGHT-QUICK RUNDOWN:

- tan/brown cap with central umbo
- wrinkling around umbo
- growing near/on rotting hardwood
- white, attached gills
- long stem that is lighter up top and darker toward the bottom
- large root present when excavated

BOLETES

Violet Gray Bolete (*Tylopilus plumbeoviolaceus*). ***This one is not edible as it's extremely bitter.**

As we touched on previously, boletes are a kind of mushroom with a spore-producing surface of pores. The teeny tiny holes on the underside of the cap are reminiscent of a sponge. What separates boletes from polypore mushrooms is that they are shaped a lot like more common gilled mushrooms. They have a cap and stem, and they fruit from the ground, but instead of gills, they have pores. Polypores are typically lacking a stem and commonly grow straight from wood.

Somethin' you may notice is that in this book, I don't include a whole lotta boletes. There's a reason for that. There are around 300 species of bolete mushrooms and identifying them can be very tricky, sometimes requiring a microscope to differentiate between certain ones. Some types have poisonous lookalikes. I don't consider boletes to be a beginner's mushroom, and I aim to keep this guidebook simple. There are some boletes that are very distinctive and edible, so I have included some of those to get you started.

Just as a quick mention, some edible boletes that can be harvested in Appalachia include, but are not limited to: porcini (*Boletus edulis*, but I've never found one personally—it's on my bucket list), chestnut bolete (*Gyroporus smithii*), Frost's bolete (*Exsudoporus frostii*), shaggy-stalked bolete (*Austroboletus betula*), two-colored bolete (*Baorangia bicolor*), ornate-stalked bolete (*Retiboletus ornatipes*), butter-foot bolete (*Boletus auripes*), old man of the woods (*Strobilomyces floccopus*), slippery jack (*Suillus luteus*), and the painted bolete (*Suillus spraguei*).

With that being said, *please* make sure you work your way up to boletes if you're just getting started with mushroom huntin'. Don't trust mushroom identification apps or a random stranger on the Internet. Always confirm with complete certainty with a reliable source.

FROST'S BOLETE *(EXSUDOPORUS FROSTII)*

There's a sayin' in the mushroom world that if you eat boletes with pores of red, it'll put you to bed, (aka kill ya). Some have also said not to eat a bolete that stains blue, buuut, neither are true here with the Frost's bolete. It sports beautiful red pores, and if you cut or scrape it, it stains a purdy dark blue and it *is* edible. Some call them candy apple boletes due to their downright gorgeous deep-red hues. One thing that rings true across the board is that this mushroom sure is easy on the eyes.

Now, I will stress, this mushroom can be confused with some other red-pored boletes that are *not* edible. The Frost's bolete's iconic features of the shiny red cap and deeply reticulated red stem set it apart, making it an easier bolete to work with. *But*, always make a bajillion percent sure your ID is correct before eating! I don't consider the Frost's bolete to be a beginner's mushroom, but it is something you can work your way up to.

This lil' frosty shows its deeply reticulated stem with those deep crevices and crannies.

LOOKS:

They typically grow to have a cap width of up to 6 inches (15 cm) with the cap being a deep-red color (hence the other common name *candy apple bolete*). The cap appears shiny, usually smooth, but sometimes it develops dimples on the cap and a yellow margin around the cap when more aged. When the mushroom is cut in half, the inner flesh is yellow and then stains dark blue after it's sliced. The underside of the cap has red pores that also bruise dark blue when damaged. The pores often exhibit what is called *guttation*, or tiny droplets of liquid oozing from the pores. In this mushroom, the guttation appears as a yellowish liquid. The stem can grow up to 5 to 6 inches (13 to 15 cm) long and is deeply reticulated with colors of red and yellow.

STOMPING GROUNDS:

They prefer to grow in mixed forests, most commonly occurring alongside hardwood trees, especially oak.

GROWTH PATTERNS:

They can be found growing solitarily, scattered, or gregariously from the ground.

SPORE PRINT COLOR:

olive brown

FLAVOR NOTES:

slightly sweet and lemony in taste and smell; the cap can taste somewhat acidic

POTENTIAL LOOKALIKES:

There are a couple other boletes, *Boletus flammans* and *Boletus rubroflammeus*, that can resemble Frost's bolete, and these are belly no-nos. These are poisonous and cause GI issues. The difference in these as compared to the Frost's bolete is that they have a much more shallow, delicate reticulation on their stem. The reticulation will not be as deep and coarse like Mr. Frost's. *Boletus flammens* likes to hang out with conifer trees, which is another way to separate it from the Frost's bolete.

RIGHT-QUICK RUNDOWN:

- shiny, candy-apple-red cap
- red pores
- reticulated stem
- bruises blue when cut/damaged
- often has guttation

CROWN-TIPPED CORAL (ARTOMYCES PYXIDATUS)

Coral mushrooms are some pretty dang cool-lookin' mushrooms. Just as the name suggests, they look like something you'd find growin' on the ocean floor amongst the coral reefs. Mushrooms seriously are so wild and unpredictable in their looks, like this freaky sea creature fella.

You'll find crown-tipped coral growing all over the place in the summer months directly from wood, which is a key factor in identification. It *can* sometimes appear to be growing from the ground, but it will actually be sproutin' from buried wood. Another key feature is in the tippy tops of their wicked lil' branches. The crown-tipped coral sports three to six prongs at the ends, so make sure you give it a real close look.

There's another type of coral mushrooms called *Ramaria* mushrooms that have a similar structure and appearance to crown-tipped coral mushrooms. The main difference in these is that you'll find *Ramaria* poppin' from the ground instead of on wood; they have a brighter range of colors than the bland, whitish crown-tipped coral; and *Ramaria* are often thicker and sturdier than crown-tipped. There are both edible and inedible varieties of *Ramaria*, so as always, make sure ya know what ya got if you stumble across a crown-tipped lookalike.

When it comes to incorporating crown-tipped coral into your kitchen adventures, I admit it can be kinda time-consuming. As you can imagine by looking at the pictures in this here mushroom profile, they can be a little bit of a booger to clean up. They have a lot of small branches, nooks, and crannies to scrub-a-dub the dirt and debris off of. They are also brittle things, so you gotta handle 'em with care to preserve the integrity of the lovely coral. When I bring a haul home, I like to fill my sink up with some cool water, break the coral up into smaller frond sections, and dunk and gently shake them around in the water to knock off the dirty bits. I then lay them out on a paper towel–lined pan to let them dry off some before throwing 'em into my food. Since these mushrooms are so delicate and mild in flavor, I like to toss them into soups and stocks or roast them whole with some olive oil and tasty seasonings atop. If you sauté them, they will likely break apart and lose that beautiful coral shape.

LOOKS:

Crown-tipped coral mushrooms typically grow to be up to 6 inches (15 cm) tall as a coral-shaped structure, including multiple branches shooting upward with three to six prongs or "crown tips" at the top of each branch. They are often white to dingy white, tanning/yellowing more with age. The branches are firm, but the overall mushroom is quite thin and brittle in texture.

STOMPING GROUNDS:

They prefer to grow directly from downed/decaying hardwood trees and logs, especially on oak, maple, tulip poplar, and aspen trees. They *can* appear to be growing from the ground at times, but they are really growing from buried wood.

GROWTH PATTERNS:

They can be found growing solitarily or in small clusters on dead/dying hardwood.

SPORE PRINT COLOR:

white

FLAVOR NOTES:

mild, kinda woody and peppery

POTENTIAL LOOKALIKES:

Strict-branch coral (*Ramaria stricta*) is another coral type mushroom that grows directly from wood. The way to differentiate these from crown-tipped coral is strict-branch coral mushrooms lack the multiple prongs, or crown tips, and they grow in a "strictly" straight up and down pattern.

RIGHT-QUICK RUNDOWN:

- coral-shaped structure with three to six prongs on each branch
- off-white to pale yellow, lighter brown when mature
- growing from hardwood

This close-up of crown-tipped coral shows the tiny prongs, or crown tips, that you're lookin' for when trying to confirm your ID of this yummy lil' guy.

CHAPTER 7

Edible Mushrooms of the Fall

Fret not, little forager! Just because the summer has ended, it does in *no* way mean that your mushroom huntin' is over for the year. Remember, I done told ya to be on the lookout for mushies year-round, and fall mushrooms happen to be some of the best and tastiest mushrooms up for grabs. It also happens to be the time of year I can really load up on some biiig hauls and get my preserving game on. Yeehaw! So, throw on a light jacket and some thicker socks, honey. Get to stompin' them woods in search of edible fungi because we ain't ever done around here, buddies!

Scan here to check out some mushrooms of the fall time!

On this fine November day, I raked in over 30 pounds (14 kg) of oyster mushrooms in one hunt and that's with leaving plenty behind! It was definitely one for the books. Fall foragin' is some of the best foragin'.

LEFT | Colorful fall leaves bring with them colorful and tasty wild mushrooms to be had.

OYSTERS (*PLEUROTUS OSTREATUS*)

When a cluster of oysters are fruiting from the top of a log, the caps can have a more rounded shape versus the fan shape they often exhibit when growing from the side of wood.

Behold one of the most abundant, commonly foraged, and tastiest mushrooms—the coveted oyster mushroom! This type of oyster occurs throughout multiple seasons and is much more common than the summer oyster mushroom we previously discussed. I find most of my *Pluerotus ostreatus* in the fall and well into the winter each year. If you know of a forest with lots of downed or dead/dying hardwood trees, that's gonna be a hot spot for these mushies.

Oyster mushrooms have gained popularity over the last couple years, and I've noticed them poppin' up on menus at restaurants all over the place. With their thick, meaty texture and mild canvas, they make for a perfect mushroom to experiment and have fun with in the kitchen.

Make sure ya pick fresh, younger specimens when ya can. The bigger and wavier the caps and the yellower the underside of an oyster gets, the funkier and more bitter the flavor becomes. A good way to evaluate an oyster's freshness is to look to the gills. The gills should be white to light cream when it is good for eatin', but they become yellow and dingy when the oyster is older and less tasty. Another thing is to give it a sniff test. Does it smell funky? You probably don't wanna eat it. Does it smell mild, mushroomy, with a hint of anise? *Eat it*!

This is also a mushroom that tends to be kinda buggy. There are these little beetles that looove oyster mushrooms called the *pleasing fungus beetle*. They are little black bugs with yellowish-orangey spots on them, and boy, do they love them some *Pleurotus* mushrooms. Don't be scurred, though. These aren't gonna bite ya, and they aren't harmful. If I find a cluster of oysters that is absolutely infested with these beetles and it's stinky and yellow, I'm gonna pass it up. If there's a few critters on the gills and I can easily brush or blow them off, I'm gonna take it home.

Scan here to see just how big of an oyster haul is possible in the wintertime!

Oysters can vary a lot in their ranges of size and color depending on the maturation. Here, we have a chunkier, heftier cluster clingin' onto a dead/dying hardwood tree.

LOOKS:

Each cap within a shelving cluster typically grows to have a cap width of anywhere from to 2 to 12 inches (2 to 30.5 cm) across, sometimes even bigger! The caps are fan-shaped and range from cream-colored to dark brown, with margins curling and becoming wavy when older. The underside reveals white to cream-colored, closely packed gills, turning more yellow with age. They have decurrent gills that run down an often off-center, stubby stem (if present). Their flesh is thick and white inside when sliced.

STOMPING GROUNDS:

They prefer to grow on living and dead/dying hardwood trees in shady areas, especially on oak, beech, and ash.

GROWTH PATTERNS:

They can be found growing in overlapping shelves or clusters directly from hardwood and seldom from conifers. They can grow on both dead and living trees, but I find mine mostly on dead/dying trees.

SPORE PRINT COLOR:

white to pale lilac

FLAVOR NOTES:

mild, slightly sweet, faint smell/flavor of anise

POTENTIAL LOOKALIKES:

Oyster mushrooms have a couple of lookalikes. *Lentinellus ursinus*, or the bear Lentinus mushroom, can resemble a cluster of oyster mushrooms. The main differences are that bear Lentinus completely lacks a stem, and it has a velvety/finely hairy cap top that has a darker brown patch near where it attaches to the wood. This dude ain't edible and is very bitter.

Another lookalike is the mock oyster (*Phyllotopsis nidulans*). The main differences here are that the mock oyster has orange-colored gills, a spore print of pink to brown, and it smells like nasty, rotten eggs.

RIGHT-QUICK RUNDOWN:

- cream to brown caps
- decurrent gills
- stubby, off-center stem (sometimes absent)
- growing in overlapping shelves from wood
- smell of anise

CREAM OF WILD MUSHROOM SOUP

INGREDIENTS

- 2 handfuls (1 ounce, or 28 g) dried mushrooms (I like to use 1 handful of morels and 1 handful of black trumpets.)
- 3 leeks
- 6 tablespoons (85 g) butter, divided
- Drizzle of extra-virgin olive oil
- 6 cloves minced garlic
- 3 cups (258 g) fresh, clean, chopped oyster mushrooms
- 2 cups (108 g) previously sautéed and frozen golden chanterelle mushrooms
- ⅓ cup (80 ml) sherry
- ½ cup (63 g) all-purpose flour
- 4 cups (946 ml) chicken or vegetable stock
- 1 cup (235 ml) mushroom-soaking liquid (the water you used to rehydrate the dried mushrooms)
- Salt and pepper to taste
- 1 tablespoon (12 g) porcini powder
- 2 teaspoons fresh thyme
- ½ cup (120 ml) heavy cream
- Bloop of truffle oil

DIRECTIONS

1. Add the dried mushrooms to a heatproof bowl and cover with boiling hot water. Allow to soak for at least 1 hour to rehydrate.
2. Cut your leeks into thin half-moon slices and wash thoroughly.
3. Melt 3 tablespoons (42 g) of butter and a drizzle of olive oil in a heavy-bottomed soup pot on medium-high heat.
4. Add the chopped leeks and cook for 5 minutes.
5. Add the minced garlic and cook for 2 more minutes.
6. Add the chopped fresh oyster mushrooms and rehydrated dried mushrooms (save the water you soaked them in!) and cook for 10 minutes until the edges start to brown.
7. Add in the thawed previously sautéed and frozen chanterelle mushrooms and cook for 2 more minutes.
8. Deglaze your pot with sherry and scrape up any browned goodies at the bottom.
9. Add in 3 tablespoons (42 g) of butter and let it melt. Add the flour and cook for 2 minutes, stirring constantly, until well combined and gloopy.
10. Pour in 4 cups (946 ml) of stock and 1 cup (235 ml) of the reserved mushroom-soaking liquid you used to rehydrate your dried mushrooms. Keep stirring as it heats and thickens up.
11. Season with salt, pepper, porcini powder, and thyme. Bring to a boil and lower the heat. Let it simmer covered for 20 minutes.
12. Remove from the heat and stir in the heavy cream and a bloop of truffle oil.
13. Serve with some crusty bread for dipping and enjoy!

Notes: I use a mix of dried, fresh, and frozen wild mushrooms. These can be adjusted to your liking and your fungi availability. Get creative and use whatever mushrooms your little heart desires! Play around with the ratio of mushrooms if you do not have enough fresh, frozen, and dried at the ready. You can do all fresh or use just frozen—it's all good! If not using dried mushrooms, replace the cup (235 ml) of mushroom-soaking liquid with an extra cup (235 ml) of stock.

Scan here to check out my delish wild mushroom soup!

HEN OF THE WOODS (*GRIFOLA FRONDOSA*)

Now *this* is a tasty mushroom, buddies! In my opinion, hen of the woods is *the* tastiest mushroom out there to get your paws on. I consider hen of the woods to kind of be like the morels of the fall. They pop up for a short window and everybody and their uncle is gonna be trying to track 'em down. They have a tongue-slappin' meaty texture and are a powerhouse of umami flavor. An added bonus is that they can get massive! Some people also call 'em *maitake* or *sheepshead mushrooms*. Not only are they a yummy mushroom, but they are also a medicinal mushroom. Hen of the woods is believed to have anti-cancer properties and is used for brain health, gut health, diabetes, and is said to be antibacterial and antiviral!

This mushroom is reaaal good at blendin' into its surroundings, so it takes a keen eye and some careful observation to spot it, another thing that makes it morel-like to me. With it's brownish-gray cap color, it can almost disappear against the leaf litter that often surrounds it. There have been times I could have sworn there was just a pile of leaves against the base of a tree and nope! It was a hen of the woods. Take your time and get close!

Hen of the woods is fickle in that it can be hard to catch in its top edible prime. I don't know how many times I've discovered a big honker hen, but it's full of bugs and it's stinky. There are these little critters called *springtails* that can hide in the pores of a mushroom. They are kinda whitish to almost translucent and are barely visible with the naked eye. You won't really even know they are there until you get it home and they start jumpin' all over your kitchen. Once I put some mushrooms in the dehydrator and went back to check and the thing was nothing but springtails! You live and you learn. They will also start trying to jump ship when you soak a mushroom, buuuut, ya already know this girl doesn't soak her wild mushrooms.

When it comes to knowing if your hen is good or not, I like to look at the pores. If the pores are small and not super visible, it's really good! If the pores have started to open up wide and you see some creepy crawlers poppin' out like whack-a-mole, leave it be.

I'm gonna let you in on a little secret of where/how I find my hen of the woods each fall. I look for big, "old property line" oaks with barbed wire fence running through them. Think about old farm roads and farmland.

Look to the base of these trees for flushes. Every single hen I have ever found has been on one of 'em! I also especially like to go on a hunt a couple of days after a big rain in September and the earlier part of October. Remember to mark your spots, too. Hen of the woods is one of those mushrooms that typically fruits back in similar spots the next year.

Hen of the woods are polypore mushrooms, so the underside of the caps will have teeny holes.

LOOKS:

The entire fruiting body can grow to be up to 3 feet (91.5 cm) wide, with the singular caps growing to be up to 3 inches (7.5 cm) wide. It grows as a dense, circular cluster of overlapping caps from a central small and thick stem. The caps are gray to brownish in color, sometimes exhibiting a range of both or faint zonations of these colors on the cap. The caps can be fan to spoon-shaped. The underside is white to off-white with small pores.

STOMPING GROUNDS:

They prefer to grow at the base of hardwood trees, especially oaks and maples, but also sometimes beech, birch, and cherry. They like to grow from trees around riverbanks and creeks.

GROWTH PATTERNS:

They can be found growing in a dense, circular clusters from the base of hardwood trees and stumps.

SPORE PRINT COLOR:

white

FLAVOR NOTES:

rich umami, woodsy, nutty

POTENTIAL LOOKALIKES:

The black-staining polypore (*Meripilus sumstinei*) can be a lookalike for hen of the woods, but the black-staining polypore has larger fan-shaped caps and also stains black when damaged, unlike hen. The black-staining polypore is edible (see summer chapter).

Another lookalike is the umbrella polypore (*Polyporus umbellatus*). The difference is that the umbrella polypore, first off, is *super* rare (it's a bucket list mushroom for me), the caps are more rounded in shape, and the caps have a more yellowish color to them. It's also edible!

RIGHT-QUICK RUNDOWN:

- dense circular cluster of overlapping caps
- single fan-shaped caps are brown to gray
- white porous underside
- growing at the base of a hardwood tree

HEN OF THE WOODS FLATBREAD

INGREDIENTS
- 1 head garlic
- 1 cup (150 g) ricotta cheese
- Salt and pepper to taste
- 1 extra-large onion
- 1 tablespoon (14 g) butter
- 1 small package (8 ounces or 225 g) fresh spinach
- Olive oil
- 4 cups (280 g) clean, chopped hen of the woods mushroom
- Flatbread (or make your own fresh)
- 1 cup (80 g) shredded Parmesan cheese
- Chopped fresh rosemary for garnish

DIRECTIONS
1. Cut the top ¼ inch (6 mm) off of a head of garlic and wrap in aluminum foil. Roast in oven at 400°F (230°C, or gas mark 8) for 1 hour.
2. Remove the garlic from the oven, peel off the aluminum foil, and let cool. Squeeze out the cloves into a bowl and mash into a paste. Mix in ricotta cheese, season with salt and pepper. Cover and let sit in the fridge.
3. Cut the onion into thin slices and caramelize in a skillet with 1 tablespoon (14 g) of butter on medium-low heat, stirring constantly, until very soft and brown (this usually takes about 15 minutes). Remove from the heat and set aside.
4. Sauté the spinach in a skillet with a drizzle of olive oil on medium-high heat until wilted.
5. Add the hen of the woods mushrooms to the skillet and sauté until tender and golden brown.
6. Add the caramelized onions to the skillet and stir to incorporate.
7. Place your flatbread on a baking sheet, spread the ricotta cheese mixture on top, and then add the mushrooms, spinach, and onion mixture. Top with Parmesan and rosemary.
8. Bake at 350°F (180°C, or gas mark 4) until the cheese has melted and the ricotta begins to brown.

LION'S MANE "CRAB" CAKES

(Lion's mane mushroom profile coming on the next page)

INGREDIENTS
- 10 ounces (285 g) lion's mane mushroom
- 6 cloves garlic
- Olive oil
- ¼ cup (40 g) diced onion
- 1 farm-fresh egg
- 1 tablespoon (14 g) mayonnaise
- 1 teaspoon Worcestershire sauce
- 1 teaspoon Old Bay Seasoning
- 2 teaspoons dried parsley
- 1 teaspoon Dijon mustard
- Salt and pepper to taste
- Juice of half a lemon
- ½ cup (60 g) breadcrumbs

DIRECTIONS
1. Clean and dry the lion's mane mushroom and shred into bite-size pieces.
2. Finely dice the garlic.
3. Sauté the lion's mane in olive oil in a skillet on medium-high heat for approximately 10 minutes or until the edges start to brown. Remove from the heat and set aside and let cool.
4. Add a drizzle of olive oil to skillet and sauté the onion and garlic for 5 minutes on medium-high heat. Remove from the heat, set aside, and let cool.
5. In a bowl, combine the egg, mayonnaise, Worcestershire sauce, Old Bay Seasoning, parsley, mustard, salt and pepper, and lemon juice until well incorporated.
6. Add the cooked lion's mane and sautéed onions and garlic to the egg mixture and stir.
7. Stir in the breadcrumbs.
8. Shape into balls (whatever size you'd like) and add to a skillet with a drizzle of olive oil on medium-high heat.
9. Flatten the balls into cakes and cook approximately 2 to 3 minutes per side until golden brown.

LION'S MANE (*HERICIUM ERINACEUS*)

Scan here to see one of the best lion's mane I've ever scored!

I sure do love a toothy mushroom, and this'un just might be the most popular and most desired one out there. Behold the famed lion's mane mushroom. It's highly sought after not only for its delicious taste, but also for its medicinal properties.

Lion's mane is such a neat stinkin' mushroom in that it has a nice mild, earthy, and sweet seafoodlike flavor, and it even has a similar texture. It's a real nice substitute in recipes that call for crab, scallops, or lobster.

As for its medicinal value, it's a cognitive powerhouse. It's been said to help with memory, focus, attention, cognition, anxiety, depression, nervous system injuries, and is often used to protect against and ease symptoms of dementia and Alzheimer's. In addition to all the brain benefits, lion's mane can also help with diabetes, ulcers, immunity, heart health, and cancer.

This one is another easy beginner's mushroom because—well, just take a gander at it! It has some very distinctive features that are hard to mistake. All its potential lookalikes are also in the *Hericium* family and are choice edibles with similar flavor profiles. It's easy to spot in the wild. It's easy to clean because it hangs from a tree and doesn't roll around in the dirt. It's meaty. It's versatile. It's frickin' gorgeous. Lion's mane has it all—beauty *and* brainpower!

LOOKS:

The whole fruiting body typically grows to be anywhere from 2 to 12 inches (5 to 30.5 cm) across, but as always, sometimes ya can find 'em even bigger (or teenier). They grow as a roundish, white to off-white mass with downward dangling spines. Some compare this mushroom to the look of a pom-pom. The mushroom is usually moist/juicy and if given a squeeze, will probably release some drops of moisture when fresh. The spines start out short and grow longer with age. The spines usually measure anywhere from ¼ inch to 2 inches (6 mm to 5 cm) long. The flesh is thick and white, and it is mostly solid throughout when sliced into a cross section. It lacks a prominent stem and is attached by its base directly to wood. The fruiting body starts to yellow with age.

STOMPING GROUNDS:

They prefer to grow on hardwood trees and on dead logs/stumps, especially growing on walnut, beech, maple, and birch. They often fruit from the wounds of trees, meaning any cracks, holes, or gashes the tree may have.

GROWTH PATTERNS:

They can be found typically growing solitarily from hardwood trees, but I have gotten lucky and found two on the same tree once before!

SPORE PRINT COLOR:

white

FLAVOR NOTES:

mildly earthy, kinda sweet, slight seafood taste

This lion's mane is one I look for every fall. Note that it's growing from the wound of the tree as lion's manes often do.

POTENTIAL LOOKALIKES:

Lion's manes have a couple edible cousins that could be potential lookalikes and those are bear's head tooth (*Hericium americanum*) and coral tooth fungus (*Hericium coralloides*). All of these *Hericium* mushrooms, including lion's mane, have white fruiting bodies with spines or teeth, but they have different patterns of growth or lengths of spines differentiating them. Here's a little breakdown:

1. **Lion's mane** will grow as one singular mass with dangling spines.
2. **Bear's heads tooth** will have multi-level branches of dangling spines.
3. **Coral tooth fungus** will have many branches with smaller spines that typically don't dangle.

Bear's head tooth and coral tooth fungus are up next so ya can really compare and see the differences.

RIGHT-QUICK RUNDOWN:

- round, white fruiting body resembling a "pom pom"
- downward dangling spines or teeth
- lacking prominent stem
- growing from hardwood trees (especially from tree wounds)

BEAR'S HEAD TOOTH (HERICIUM AMERICANUM)

Since we are on the toothy topic, here's another spiney feller: bear's head tooth. Just like lion's mane, it's gonna be found growin' in similar habitats and conditions and will have that lovely, mildly sweet seafood taste. The main difference here, as we discussed in the previous profile, is the growth pattern. Bear's head tooth has multiple branches and levels of dangling spines versus the rounded, single mass of the lion's mane. I've heard some folks call bear's head tooth lion's mane before, which reinforces the importance of those Latin binomials—so we know exactly what each other are talkin' about.

These branching spines are a dead giveaway to knowin' ya got a bear's head tooth.

LOOKS:

The whole fruiting body typically grows to be anywhere from 6 to 12 inches (15 to 30.5 cm) across and up to 10 to 20 inches (25.5 to 51 cm) tall. When fresh, the mushroom is white throughout and often moist/juicy, yellowing with age. It exhibits groups or multi-levels of branches of dangling spines/teeth that hang downward, with the spines measuring usually a ½ inch to 1 inch (6 mm to 2.5 cm) long. The mushroom lacks a prominent stem and is attached to wood by a thick base.

STOMPING GROUNDS:

They prefer to grow from the trunk, stumps, or logs of living or dead/dying hardwood trees, especially beech, oak, birch, walnut, and sycamore. I find most of mine on dead/dying hardwood.

GROWTH PATTERNS:

They can be found growing solitarily or in small groups directly from wood.

SPORE PRINT COLOR:

white

FLAVOR NOTES:

mildly earthy, kinda sweet, slight seafood taste

POTENTIAL LOOKALIKES:

Lion's mane (*Hericium erinaceus*) and coral tooth fungus (*Hericium coralloides*) are lookalikes, but edible. See the lion's mane profile lookalikes for a rundown on these again and see the next profile for more information on coral tooth fungus.

RIGHT-QUICK RUNDOWN:

- white fruiting body with multilevel branching
- long, dangling white spines
- growing from hardwood
- thick base attaching to wood (no prominent stem)

CORAL TOOTH FUNGUS (HERICIUM CORALLOIDES)

And now for the third toothy member of the *Hericium* mushroom family, the coral tooth fungus. This guy isn't as showy with his pearly white teeth as the lion's mane and bear's head tooth, but he's got some teeth, nonetheless. You may also hear this mushroom called a *comb tooth mushroom*, which is a good name because the arrangement of the teeth on the branches can also resemble the teeth of a comb. Just like the others, this one lends that yummy, sweet, mushroomy, seafoody taste.

The name lets ya know—this mushroom looks an awful lot like coral. The cluster appears directly from wood and is white, branchy, and usually has shorter lil' teeth than the two *Hericium* previously discussed. Given the shorter length of the teeth, they don't necessarily dangle as much as lion's mane or bear's head tooth.

Note the much shorter teeth here of the coral tooth fungus as compared to lion's mane and bear's head tooth. Another dead giveaway to me are the gaps and spaces in between the branches.

LOOKS:

The whole fruiting body typically grows to be up 12 inches (30.5 cm) across and 10 inches (25.5 cm) high. The mushroom is white throughout, yellowing with age. Its shape and appearance are reminiscent of coral (hence the name) and has many branches shooting out in all directions, with visible gaps and space in between the branches. When young, the mushroom may appear branchy and without teeth, but the teeth can grow in later. Each branch has teeth on it that are arranged in lines and those teeth commonly measure 0.1 inches to 0.4 inches (2.5 to 10 mm) long. It lacks a prominent stem and attaches from a base directly to wood.

STOMPING GROUNDS:

They prefer to grow from the trunk, stumps, or logs of living or dead/dying hardwood trees, especially beech, oak, birch, walnut, and sycamore. I find most of mine on dead/dying hardwood.

GROWTH PATTERNS:

They can be found growing solitarily or in small groups directly from wood.

SPORE PRINT COLOR:

white

FLAVOR NOTES:

mildly earthy, kinda sweet, slight seafood taste

POTENTIAL LOOKALIKES:

Lion's mane (*Hericium erinaceus*) and bear's head tooth (*Hericium americanum*) are lookalikes, but edible. See the lion's mane profile lookalikes for a rundown on these again.

RIGHT-QUICK RUNDOWN:

- white fruiting body with multiple branches
- visible gaps/space between branches
- looks like coral
- short teeth
- growing directly from wood

WOOD BLEWIT (COLLYBIA NUDA)

Sometimes, Mama Nature can pump out some pretty cool and mesmerizin' mushrooms, and this here certainly is one. Don't be scurred of its vibrant coloring. Wood blewits are a totally safe, tasty, purdy, purple forest treat. Say that ten times fast. Remember, there's no general or blanket rule to determine whether a mushroom is edible or not, so never go by color to make that decision.

This is one of those mushrooms that when ya see one, there's probably gonna be more right close. They like to grow in a scattered pattern, which we know means that mushrooms of the same species will be growin' usually within 1 foot (30.5 cm) or less of each other. They are easy to spot because they are an obvious purple pop usually amongst more earth-toned leaf litter or woody debris, so she stands out for sure. This mushroom can also be found outside of the woods, such as in your compost pile or even your own yard.

A fun thing this mushroom has going on is its smell. If you get'cha a pull of a wood blewit with your nose holes, it smells sweet and fruity, kinda like a mix of orange juice and lilac flowers. This is a good way to tell them apart from potential toxic lookalikes.

As with any new-to-you wild mushroom, eat only a small portion and cook thoroughly to get started. Also make sure you're eating a healthy, younger specimen, as the older the mushroom, the less flavorful and the more likely it is to give you bubble guts. Wood blewits are a mushroom that has been known to not agree with some people's tummies.

Wood blewits will dull in color as they mature, as seen here on the cap of the left mushroom.

LOOKS:

They typically grow to be anywhere from 2 to 6 inches (5 to 15 cm) wide across the cap with a stem that is up to 4 inches (10 cm) long. The cap is darker purple at first when young. As it ages, it turns to a lighter lilac and will eventually become a light-brown or tannish color when very mature. The cap is smooth, starting out button-shaped, flattening, and then becoming wavy at the cap margins with age. The underside of the cap has closely spaced, purple gills that become browner with age. The stem is often lighter purple in color than the cap, and the stem can sometimes have a fatter base to it. The stem has no ring or partial veil. The flesh inside is light purple to off-white, depending on age, when doing a cross section.

STOMPING GROUNDS:

They mostly prefer to grow in leaf litter underneath oak trees in my experience, but they can also be found growing in woody debris, gardens, compost, and lawns. They can grow in mixed woodland areas, so really just be on the lookout anywhere for these.

GROWTH PATTERNS:

They can be found growing sometimes solitarily, but mostly scattered to gregariously from the ground.

SPORE PRINT COLOR:

off-white to pale pink

FLAVOR NOTES:

slightly sweet and peppery

POTENTIAL LOOKALIKES:

Some *Cortinarius* mushrooms are a common lookalike for the wood blewit. The major differences of mushrooms in the *Cortinarius* family are that they have a rusty brown spore print instead of the off-white to pale pink spore print of the wood blewit. Oftentimes, you will be able to see the brown spores on the stem or around the outer part of the cap. *Cortinarius* also have a fibrous partial veil that looks like a spider web, whereas wood blewits have no ring or partial veil. Also, remember that wood blewits have that sweet, fruity smell and *Cortinarius* often smell like radish. Many *Cortinarius* mushrooms are poisonus, so don't eat 'em!

RIGHT-QUICK RUNDOWN:

- smooth, purple cap
- closely spaced purple gills
- light-purple stem with no ring or partial veil
- growing scattered from the ground near hardwood trees (especially oak)
- smells of orange juice and flowers

HONEY MUSHROOMS

This is a cluster of some dated ringed honey mushrooms (*Armillaria mellea*). In their older age, the margins of honey mushroom caps will upturn.

You'll often find ringless honey mushrooms (*Desarmillaria caespitosa*) growing in tightly packed clusters from the ground.

Honey mushrooms are incredibly common in the fall and there's more than one type of mushie people call "honey mushrooms." The ones I seem to see the most are *Armillaria mellea* and *Desarmillaria caespitosa* (formerly *Armillaria tabescens*), or ringed and ringless honey mushrooms, respectively. They appear as a tightly packed cluster of honey-colored mushrooms, ranging from yellow to brown, and they either do or do not have a ring on their stem. The caps typically have a darker spot in the middle, and the gills are whiteish. They are usually on/near the base of wood (see photos) or growing from buried wood. So, these *are* edible mushrooms, however, I'm not a huge fan of their texture/flavor. They tend to have slimy caps and can be bitter if not very young. They are also a mushroom that has caused gastric woes with a number of folks. If you do eat them, it's recommended to boil them for 10 minutes to break down anything that's gonna rock your gut. Given these reasons, I'm not including a full profile on these type of honey mushrooms, but I will talk about another feller, the bulbous honey fungus, especially since it plays a key role in another type of edible mushroom later, shrimp of the woods.

BULBOUS HONEY FUNGUS
(ARMILLARIA GALLICA)

These here are some honey fungi with some junk in the trunk, hence the name *bulbous* honey fungus. They are a fun lookin', plentiful mushroom with a slightly sweet smell. My personal favorite feature I find to be the coolest on this mushroom would be the freaky lil' cobwebs they usually wear under their hood. As we learned previously, some mushrooms have a partial veil in their younger age that protects their spore-producing surface. The bulbous honey fungus has just that. It eventually grows up and flattens out, leaving the partial veil remnants around the stem. These remnants look very wispy and cotton candy–like, making this a surefire way to identify your honeys.

Just as with the honey mushrooms mentioned previously, you're gonna wanna make sure you cook the heck out of these in order to remove that bitter/acrid taste they can have when not cooked properly. Another important reason to do this is because it can be hard on the belly for some. Try just a small portion to see how it sits with ya. The taste is not gonna blow your socks off, but they can be dressed up and be a welcomed addition to your plate. The main reason I am including bulbous honey fungus is because of its tastier alter ego, shrimp of the woods (see next profile).

Note the cobweb-like material that many bulbous honey fungi show when you flip 'em over. It's a major identification feature for these guys.

Here's a bulbous boy who is fruitin' directly from wood. He's showing off his shaggy stem, bulbous base, and remnants of his partial veil on the stipe.

LOOKS:

They typically grow to have a cap width of 1 to 5 inches (2.5 to 13 cm) wide, and the whole fruiting body usually stands 2 to 5 inches (5 to 13 cm) tall. The caps range from pinkish tan to yellowish brown in color and sometimes exhibit small bumps or scales and sometimes fine, white hairs. When young, the cap is more rounded, flattening with age. The cap will oftentimes have whitish to yellow remnants around the margin from its partial veil. The gills are white to tannish pink in color and are frequently adnate and subdistant. The stem usually appears shaggy and mostly light-colored, but sometimes has streaks of pinkish brown. The stem is often larger or "bulbous" at the base. When young, the mushrooms have a partial veil, and when older, they eventually have a ring on the stem with cobwebby remnants present near the gills/ring.

STOMPING GROUNDS:

They prefer to grow from buried roots near the base of hardwood trees, stumps, and fallen logs. Sometimes, they fruit directly from dead/dying hardwood. I also tend to find them around moss.

GROWTH PATTERNS:

They can be found growing solitarily, but most often are found growing scattered or in clusters either near decaying wood or directly from dead/dying wood. They rarely grow from conifers.

SPORE PRINT COLOR:

white to cream

FLAVOR NOTES:

mild to bitter, slightly sweet

POTENTIAL LOOKALIKES:

Shaggy pholiota (*Pholiota squarrosoides*), which is not edible, could be confused for bulbous honey fungus at first glance. The main differences here are that shaggy pholiota will produce a brown spore print and they have more three-dimensional, very obvious spikey scales on both the cap and stem. Bulbous honey fungus have a white spore print and smaller, less prominent scales on the cap only.

RIGHT-QUICK RUNDOWN:

- pinkish-tan to brown caps with small scales or fine hairs
- white to pinkish-tan gills
- shaggy stem
- bulbous base
- partial veil when young, with ring on the stem later with cobwebby remnants

SHRIMP OF THE WOODS (ENTOLOMA ABORTIVUM)

Okay, I love me some shrimp of the woods. They look so strange, but they are so good. When found in peak stages, I could easily eat it by the bucket full.

These white deformed balls of goodness often appear like little popcorn balls or heads of garlic sproutin' out of the ground at the base of trees. The process in which they come to be is really dang neat. You'll also hear them called *aborted entolomas*. In the mycology world, they kind of did a back-and-forth over what mushroom caused the other mushroom to produce shrimp of the woods. I think they finally figured it out, so I'm gonna share with you the wild process that makes these tasty globs.

There's a mushroom called the *Entoloma* mushroom, and if it grows near honey mushrooms (often bulbous honey fungus, see previous profile, page 143), it can parasitize the mycelium of the honey mushroom. This causes the honey mushroom to grow all weird and wonky, resulting in shrimp of the woods. Calling them aborted entolomas isn't necessarily accurate since it's *actually* the honey fungi that is aborted. Both the aborted forms and the gilled forms of these mushrooms are edible, but always be sure of your identification before cramming it in yer mouth.

They are called shrimp of the woods not so much for their taste, but for their texture. Another wild thing about this mushroom is it lacks any gills or pronounced stem. It's just like a weird orb of funkiness.

Finding these babies can sometimes be like a treasure hunt. If I spot one, I will shuffle the leaf litter around in the general area. Nine times outta ten, I'm gonna find more tucked away underneath. When you find a big haul of shrimp of the woods, sometimes you may even find some honey mushrooms in multiple stages: not quite aborted, kinda aborted, or completely aborted (shrimp of the woods). I opt for the completely aborted forms. I always make sure and put things back the way I found them if I find myself doing any rearrangement of leaves and whatnot.

Make sure you give these a good brushin' in the field with your mushroom brush. These are notorious for having nooks and crannies that can hold a whole lotta dirt.

As you can see here, shrimp of the woods lack any gills or stem you see in other mushrooms.

I often find clusters of shrimp in beds of moss at the base of hardwood trees.

LOOKS:

They typically grow to be anywhere from 1 to 5 inches (2.5 to 13 cm) wide and often lack prominent stems or gills. There will sometimes be a nub present that attaches the mushroom to the ground, which can be perceived as a shorty stem. They appear as rounded white to grayish lumpy balls all wadded together with a kinda dusty looking or cracking exterior. Oftentimes, they exhibit a central depression in the top of the "cap." The flesh is firm when fresh. In older specimens, the flesh becomes softer and browner in color. There can sometimes be flecks of pink inside of the mushroom when doing a cross section.

STOMPING GROUNDS:

They prefer to grow near decaying wood, moss, and where you may have seen honey mushrooms growing previously. I find a lot of mine hanging out near the base of trees and in the roots of hardwoods, especially beech, elm, and birch for me, but they can also be found in coniferous forests.

GROWTH PATTERNS:

They can be found growing most often in clusters or scattered groups, and less often you may see them growing solitarily.

SPORE PRINT COLOR:

salmon pink

FLAVOR NOTES:

mild, nutty

POTENTIAL LOOKALIKES:

Some *Amanita* species of mushrooms can start out as a white egg lookin' sac that grows from the ground. These little eggs could potentially look like shrimp of the woods, but there's a super simple way to tell what ya got. When you cut open the sac of the baby *Amanita*, you will see the outline of a mushroom inside, with lines of the mushrooms soon to be stem and gills. When you cut open a shrimp of the woods, you will not see any evidence of a mushroom about to be birthed inside of it. Remember, the *Amanita* family has a whole lotta toxic mushrooms in it, so please be careful!

RIGHT-QUICK RUNDOWN:

- popcornlike, white, dusty globs
- growing near decaying wood or at the base of trees
- cross section reveals flesh that is reminiscent of the texture of shrimp
- firm texture when fresh
- near other honey or *Entoloma* mushrooms

Scan here to see how shrimp of the woods are made and found!

POPCORN SHRIMP OF THE WOODS

INGREDIENTS
- 1 pound (455 g) shrimp of the woods mushrooms
- 1 cup (125 g) all-purpose flour
- Salt and pepper to taste
- 1 farm-fresh egg
- 1 tablespoon (15 ml) water
- ½ cup (56 g) panko breadcrumbs
- ¼ cup (30 g) plain breadcrumbs
- ½ teaspoon garlic powder
- ½ teaspoon onion powder
- Oil for frying (I use veggie oil)
- Cocktail sauce (optional)

DIRECTIONS
1. Clean the shrimp of the woods mushrooms and snip off any dirty parts that cannot be cleaned. Discard any mushrooms that are mushy or browning.
2. Set up a breading station with three bowls. To the first bowl, add the flour and salt and pepper to taste and mix. To the second bowl, add the egg and water and whisk. To the third bowl, add the panko, plain breadcrumbs, salt and pepper to taste, garlic powder, and onion powder and mix.
3. Heat enough oil for frying on medium-high heat in cast iron skillet.
4. Dredge the clean shrimp of the woods in the first bowl of flour and shake off excess, dunk in egg/water mixture, and then finally dredge in breadcrumb mixture. Set aside.
5. Once the oil is to temperature, carefully add in breaded shrimp of the woods and fry for a couple minutes per side, flipping as needed until golden brown on all sides.
6. Remove from the oil and place on a paper towel–lined pan or a rack to let excess grease drain.
7. Serve with cocktail sauce if desired and eat up!

Note: When I get a big ol' haul of shrimp of the woods that can't all be eaten in one sitting, I follow this recipe by breading them, but instead of frying them, I freeze 'em. I lay my breaded nuggets of shrimp of the woods in a single layer on a sheet pan to freeze for an hour. Then, I plop them into freezer bags and shove 'em into my foraging deep freeze. Whenever I've got a hankering for some fried shrimp of the woods, all I gotta do is heat me up some oil and fry away no matter what time of year it is!

Scan here to watch me make some popcorn shrimp of the woods!

AMBER JELLY ROLL (EXIDIA RECISA)

I'll never get tired of wigglin' every single gob of jelly fungi I find in the woods. It's basically an involuntary reflex for me when I spot 'em, so naturally, ya girl sure does love to jiggle this here amber jelly roll. It has a few different common names, including brown witches' butter and willow brain. In the fall, this stuff absolutely loves to grow and cling to branches and twigs of hardwood trees, especially beech and willow. I'll find it growing in abundance on loose branches that are dangling in the tree canopy overhead as well as ones that have detached and fallen, so make sure you're lookin' both up *and* down for it on hunts. As an added bonus, amber jelly roll can grow well into the winter, so keep them eyes peeled even in the snow!

As we've discussed before, jelly fungus needs a lil' help in the kitchen. It usually is bland in flavor, but this provides ample opportunity for you to get wild and try all kinds of things with it. It's a blank canvas that will take on whatever sauce, broth, or seasonings you cook it in. Amber jelly roll can even be made into a sweet treat. Nope. I didn't stutter. Check out the recipe below for this'un and blow everyone's mind with your sweet mushrooms!

AMBER JELLY ROLL GUMMY CANDIES

INGREDIENTS
- 2 cups (475 ml) water
- 2 cups (400 g) sugar
- 1 cup (235 ml) grape juice (You can use whatever fruit juice you like.)
- Juice of 1 lemon
- A few big handfuls (2 cups, or 198 g) of amber jelly roll mushrooms
- Extra sugar for coating

DIRECTIONS
1. Bring the water, sugar, and fruit juices to a boil on the stove.
2. Add the amber jelly roll mushrooms to the boiling mixture and let boil for about 10 minutes on medium-high heat, stirring occasionally.
3. Strain the mushrooms from the mixture.
4. Spread the mushrooms in an even layer onto dehydrator trays.
5. Dehydrate until they are at desired texture, usually around 6 to 8 hours.
6. Remove from the trays and immediately roll in sugar for coating.
7. Feed them to folks and watch them freak out when you tell 'em it's mushrooms.

Note: You can add some food grade citric acid to your last sugar coatin' mixture to create sour mushroom gummies!

Would ya just look at how much amber jelly rolls 2 branches can get ya?! On this particular day, I brought home sooo much.

Amber jelly roll can also grow in a more scattered pattern on branches instead of tight clusters, as seen here.

LOOKS:

They typically grow as globular, brainy leaflike masses and each fruiting body usually measures ¼ inch (6 mm) to up to 3 inches (7.5 cm) wide. They grow in clusters but are made of individual little mushrooms. Their color is typically brown to dark brown. Oftentimes, it exhibits a wrinkled, gelatinous appearance. The texture is smooth and squishy. They can dry up and become small, brittle, and black depending on weather conditions. If found in its dried state, a rain or a soak in water at home rehydrates it and returns it to its funky, jiggly self. It lacks a prominent stem but does have a tiny portion that attaches the mushroom to the wood it is fruiting from.

STOMPING GROUNDS:

They prefer to grow on detached sticks, branches, and twigs of hardwood trees, especially beech, oak, and willow trees. I have also seen them growing on attached branches as well.

GROWTH PATTERNS:

They can be found growing in overlapping clusters or scattered groups on wood.

SPORE PRINT COLOR:

white

FLAVOR NOTES:

doesn't have a ton of flavor and takes on the flavor of whatever sauce/broth/seasoning you cook it in

POTENTIAL LOOKALIKES:

Wood ear (*Auricularia angiospermarum*) is a common lookalike for amber jelly roll. Amber jelly roll is different in that it is usually darker brown in color, smaller in shape, is squishier and more bloblike, and you can squeeze it and easily change the shape or stretch it. Amber jelly roll does not have tiny hairs and is typically wrinklier than wood ear. Amber jelly roll likes to grow more on branches and sticks instead of on the larger pieces of wood the wood ear prefers.

RIGHT-QUICK RUNDOWN:

- brown to dark-brown, globular, jellylike masses
- wrinkled and gelatinous
- growing on sticks/twigs of hardwood trees
- squishy

SNOW FUNGUS (TREMELLA FUCIFORMIS)

Yay! More jelly! Snow fungus is so dang easy on the eyes. It grows as a gorgeous frosty white, translucent mass on wood. It can have a sort of coral-like structure or appearance to it with its multiple gelatinous fronds. What separates this jelly fungi from others is that its structure has more pronounced, upright lobes rather than bein' a big blob of shaky jelly.

Not only is snow fungus outright edible in both sweet and savory applications, but it's also a well-known medicinal mushroom. It is used extensively in Chinese medicine and is said to have anti-inflammatory, immune boosting, and antioxidant properties. Snow fungus is also used in skin care products to promote moisturization and anti-aging benefits.

Note the intricate white translucent fronds of the mesmerizing snow fungus.

LOOKS:

The whole fruiting body typically grows to be anywhere from 2 to 6 inches (5 to 15 cm) across. It has a frosty white color to it while exhibiting some translucency. It is often watery in feel and appearance. It is made up of individual fronds or lobes that reach upward, with thinner margins. The fronds are smooth and shiny and will have observable spacing in between frond folds.

STOMPING GROUNDS:

They prefer to grow directly from the wood of dead/dying hardwood trees, especially oak and maple, and occasionally on conifer trees like spruce and pine.

GROWTH PATTERNS:

They can be found growing solitarily or gregariously on hardwood.

SPORE PRINT COLOR:

white

FLAVOR NOTES:

doesn't have a ton of flavor and takes on the flavor of whatever sauce/broth/seasoning you cook it in

POTENTIAL LOOKALIKES:

White jelly fungus (*Ductifera pululahuana*) is a potential lookalike for snow fungus. The main differences are that white jelly fungus is not translucent and appears as more of a blob instead of having the gentle lobes that snow fungus has. White jelly fungus often looks more like like a ribbon of thick folds instead of having the dainty coral appearance of snow fungus.

RIGHT-QUICK RUNDOWN:

- translucent to pale white in color
- jellylike in texture (sometimes wet)
- multiple branching fronds or lobes

CAULIFLOWER MUSHROOM
(SPARASSIS AMERICANA)

Usually, I'm all supportive of common names of mushrooms, but I believe the powers that be really missed the mark when they didn't name this'un the egg noodle mushroom. Oh well, cauliflower is close enough I reckon. This is a big feller that loves to fruit at the base of conifer trees. They love to do it year after year, so mark your maps and calendars, buddies, because it's liable to be back! Also, since it grows like a big honkin' white to cream-colored mass, it sure is an easy one to spot.

Though it is tasty, it can be a pain to carry out of the woods and clean. When harvesting this clump, ya wanna make sure you grab firmly at the base that attaches it to the ground and cut the stem above ground. We like to leave the mycelium undisturbed and the "root" underground so we can enjoy that same mushroom again next year. Ya don't want it to roll around in the dirt any more than it must, so get'cha a good grip on it. I've been known to carry this mushroom out of the woods in my hands instead of throwing it in my mushroom sack because of its fragility. The little fronds of this mushroom will breaky easily, so I don't like it smacking around in my mushroom bag and turning into cauliflower shrapnel.

Cauliflower mushrooms have a lot of little gaps that dirt and debris cling to, so I find that separating the mushroom into a few pieces really helps with the cleanin' process. Much like I do with my black trumpets, I like to get a bowl of cool water and give my cauliflower mushroom chunks a good swishing until I'm pleased with my cleanin' job.

I love cooking with cauliflower mushrooms because the texture holds up nicely and you can treat it like the noodles that it resembles. I've sautéed it up and placed a protein atop of it, and it's just like you've got a nice lil' bed of egg noodles below. So good!

As with all mushrooms, give it a sniff and a feel to make sure you have a good specimen that is ripe for the pickin'. If it is yellow, mushy, or smells funky, leave 'er be.

Note the ruffly, smooth surfaced branches that make up the cauliflower mushroom.

LOOKS:

The entire fruiting body can grow to be up to 12 inches (30.5 cm) across and stand up to 12 inches (30.5 cm) tall. It shows itself as an off-white to cream-colored tightly packed cluster of ruffly, egg noodlelike branches, yellowing with age. The individual "noodles" or branches of the mushroom are thin and fragile. It has a thick central brown- to black-colored stalk that is deeply rooted into the ground at the base, anchoring it into the earth below.

STOMPING GROUNDS:

They prefer to grow at the base, near stumps, or from the buried roots of conifer trees, especially pine.

GROWTH PATTERNS:

They can be found growing most often solitarily at the base of conifers.

SPORE PRINT COLOR:

white

FLAVOR NOTES:

earthy, nutty, savory

POTENTIAL LOOKALIKES:

There's another type of edible cauliflower mushroom, *Sparassis spathulata*, that looks similar to *Sparassis americana*. I tend to find *Sparassis spathulata* less frequently than *Sparassis americana*. The main differences are that *Sparassis spathulata* likes to grow near oak trees (but sometimes conifers) and has thinner, ruffling margins, whereas *Sparassis americana* likes to fruit from conifers and has thicker lobes than the *Sparassis americana* mushroom. *Sparassis spathulata* also will sometimes have zonations and *Sparassis americana* does not.

Note the thinner ruffles of the Sparassis spathulata *as compared to* Sparassis americana.

RIGHT-QUICK RUNDOWN:

- off-white to cream in color
- tight, ruffly clusters resembling egg noodles
- thick, central, brown to black stalk
- fruiting at the base of conifer

LOBSTER MUSHROOM
(HYPOMYCES LACTIFLUORUM)

Here's another wild one, the lobster mushroom. It's kinda like how shrimp of the woods was tryin' to be another mushroom, was attacked, and turned into an even tastier mushroom. Lobster mushrooms are parasitized by the *Hypomyces* fungus. Again, the joke's on the attacker because this results in a better textured and better tasting mushroom that is the lobster mushroom.

These lobster mushrooms typically will start out as *Lactarius* (milk caps) or *Russula* mushrooms, but then here comes *Hypomyces lactifluorum* to shake things up. See, this really isn't the fruiting body, but the film that grows over the host mushroom. *Hypomyces lactifluorum* is a parasite that grows on other mushrooms; it is not the body itself. Strange, huh? This sheet of fungus that creeps all over the *Russula* or *Lactarius* completely changes the overall appearance of the mushroom, turning it into the bright deep reddish orange color as pictured here. It too results in the gills either being completely absent or more ridgelike in appearance, and the mushroom develops an overall sturdier texture and stature. The parasitization also makes the host mushroom grow to be a warped, wonky lookin' thing. They are wavy and oddly shaped and often will develop underground. You may see them as weird ground lumps in the forest, barely peepin' out. Careful where ya step!

LOBSTER MUSHROOM DUXELLES

Duxelles (pronounced duck-sells) is a mix of finely chopped up mushrooms, onions or shallots, herbs, and seasonings that is cooked down in butter. Look at this hillbilly gettin' all French and fancy. Duxelles is crazy versatile and can be incorporated into tons of dishes including pasta, eggs, spreads, sandwiches, and fillings. You can use any mushroom of your likin' to whip ya up some.

INGREDIENTS
- 1 pound (455 g) cleaned lobster mushroom, diced finely
- 3 tablespoons (42 g) butter
- 1 medium shallot, diced
- ½ tablespoon fresh thyme leaves
- 5 cloves minced garlic
- Salt and pepper to taste

Lobster mushrooms like to grow in all kinds of funky shapes and sizes.

LOOKS:

Hypomyces lactifluorum is a fungus that has covered the host mushroom. It appears as a bright orangish red film that is thick and bumpy all over the outside of the mushroom. The parasite will cause the host mushroom to lack gills completely or have very faint ridges under the cap. The stem of a mushroom parasitized is often stumpy and short, and the cap will take on odd, contorted shaping. A cross section of a mushroom that has been parasitized usually reveals white flesh inside.

STOMPING GROUNDS:

They prefer to grow around hemlock and pine trees. You can find them where you have seen milk caps and *Russula* mushrooms growing previously.

GROWTH PATTERNS:

They can be found growing solitarily or scattered from the forest floor.

SPORE PRINT COLOR:

white

FLAVOR NOTES:

earthy, nutty, slightly sweet seafood taste

POTENTIAL LOOKALIKES:

Lobster mushrooms could *maaaybe* be confused with chanterelles. They both can have a bright orange color, thick stem, and a trumpeted cap. The main difference is that chanterelles have more prominent and obvious false gills under the cap whereas lobsters will not. Both are good edibles!

RIGHT-QUICK RUNDOWN:

- bright reddish orange in color
- white flesh inside
- stubby stem
- lacking gills or has very faint ridges under cap
- growing near conifers

DIRECTIONS

1. Dry your mushrooms as much as possible.
2. Heat the butter in a skillet on medium-high heat and add the mushrooms, shallot, thyme, garlic, and salt and pepper.
3. Cook, stirring often, until the mushrooms release their liquid.
4. Continue cooking until the liquid evaporates and the mushrooms start to brown.
5. Remove from the heat and let cool.
6. Once cooled, add to a freezer-safe container and throw them in the freezer until you're ready to use.

Note: This is a great thing to make when you have more mushrooms than you know what to do with and want to preserve them. Freeze it and then you can break off chunks as needed for recipes.

RESINOUS POLYPORE
(ISCHNODERMA RESINOSUM)

This mushroom is one that if ya didn't know it was edible, ya might just breeze by it thinkin' it ain't worth a dang. Tougher-appearing bracket fungi that grow from wood often get a bad rap for being too tough and rubbery to eat. Resinous polypore is an exception! Now, you only want to harvest them when they're very young and tender. When they are prime, they have a nice, sweet smell and exude red droplets of moisture, which is how this mushroom gets its name. The droplets resemble resin. A good way to test for freshness is to give this guy a squeeze and see how tough he is. If it feels soft and has some good give and moisture to it, it's in a good edible stage. If it's super tough, hard, and dry, let it go. You wanna look for little pinkish nuggets fruitin' from the wood for top eating-quality resinous polypores, kinda like little nubs. These are best.

If you bring home a haul of these, you may find that the back end of the mushroom that clings to the wood will be woodier than the tender outer edges. These tougher bits don't have to be tossed out the winder. Instead, you could opt to dehydrate them and powder them for a mushroomy seasoning, or you could add these dried chunks to soups/stocks for an extra pop of umami flavor.

As a side note, when you come across an old and tough specimen, they sure are fun to perform drum solos on with a couple sticks if you're feelin' froggy.

Resinous polypores have small pores on their white underside, which may be harder to see in fresh, young specimens. Squint!

LOOKS:

They typically grow to be up to 12 inches (30.5 cm) wide and are fan-shaped or semicircular. When young, the top of the caps are velvety in appearance and have a pale pinkish to brown color. They look like little nubs and are quite thick at this point. You will also likely see a white margin on the outer part of the cap. As they grow and age, the caps become darker brown and tan in color with differing zonations. It likely will be darker in color near its point of attachment to the wood. The underside is white and porous, with the pores being teeny tiny. The underside often bruises brown when damaged. The spore-producing surface becomes darker with age. Depending on maturation, these mushrooms often have reddish-gold droplets of moisture that resemble resin on them.

STOMPING GROUNDS:

They prefer to grow on dead/dying hardwoods and sometimes conifer trees. I find most of mine on fallen logs and have good luck with maple, oak, and ash trees.

GROWTH PATTERNS:

They can be found growing solitarily or in overlapping groupings on wood.

SPORE PRINT COLOR:

white

FLAVOR NOTES:

earthy, mild

Here, we have a younger resinous polypore showin' off how it earned its name, with those droplets reminiscent of resin. These liquid droplets are called guttation.

POTENTIAL LOOKALIKES:

Ganoderma species of mushrooms, which we know now as reishi (see page 106), can be potential lookalikes of resinous polypore. The main differences are that reishi are typically tough/hard throughout all stages of maturation, never feeling soft or moist. Reishi also have a shinier, varnished looking cap, whereas resinous polypore does not. Many *Ganoderma* species are medicinal in teas and tinctures, but not really *edible* because you'd break your teeth out.

RIGHT-QUICK RUNDOWN:

- fan-shaped clusters on wood
- velvety appearing cap
- tan to brown zonations
- white, porous underside
- droplets of reddish-gold moisture

PURPLE-GILLED LACCARIA
(LACCARIA OCHROPURPUREA)

This gal sure is a beauty and sure is plentiful in the fall. The hillside by my house gets plumb covered with 'em. I see those buff caps scattered around as far as my eye can see. No one talks about her much, and you know I'm always rootin' for the underdog. Although she doesn't boast an insanely punchy flavor, she has a nice mild mushroomy flavor you can incorporate into dishes. Adding some purple-gilled *Laccaria* in with a wild mushroom mix is a delight. She's a welcome addition to stir-fry, roasted veggie medleys, soups, stews, and whatever else. Think anywhere you'd put those lame store-bought button mushrooms, but this is way cooler because it's a wild mushroom you picked. Remember, you can always pump up the volume on the flavors by addin' fun sauces and seasonings. Unfortunately, she won't retain her purdy purple color when cooked, just like with *Lactarius indigo*, but that's okay. You get to admire the color before consuming. Make sure you snap some pics because she's oh-so-photogenic.

When it comes to this mushroom, I do only eat the caps. I've tried the stems before, and it just really ain't cute. The stem is real stringy and fibrous, which doesn't make for a very pleasant mouthfeel. The stem is also often a cozy home for bugs, so yet another reason to pass on 'em. When I harvest, I go ahead and cut the stems off in the field and drop 'em. Remember, we like to trim and drop as much as we can! It leaves little forest nuggets for the critters to munch on.

Purple-gilled Laccaria *caps are the best part for eatin'. Note those distinct distant and rigid gills they house under their hood.*

Here, you can better see the off-white smooth cap of the purple-gilled Laccaria.

LOOKS:

They typically grow to have a cap that is up to 5 inches (13 cm) across, and they can stand up to 8 inches (20 cm) tall. When young, the cap starts out a pale purple color and eventually fades to a buff white when older. The cap is mostly smooth. The gills are rigid, waxy, and distantly spaced. The gills often attach to the stem and are a beautiful vibrant purple color but fade with age. You will note many gaps and spaces between the gills. The stem is colored similarly to the cap, ranging from light purple to buff white depending on age. The stem often has a striated or scaly appearance. The stem is usually uniform in shape, but sometimes can be a bit larger or swollen at its base.

STOMPING GROUNDS:

They prefer to grow in mixed forests from the ground. I find mine mostly around hardwood trees, especially with beech and oak, but they can also grow with conifers such as pine.

GROWTH PATTERNS:

They can be found growing solitarily, scattered, or gregariously from the forest floor.

SPORE PRINT COLOR:

white to pale purple

FLAVOR NOTES:

mild, earthy

POTENTIAL LOOKALIKES:

Some *Cortinarius* species of mushrooms can resemble purple-gilled *Laccaria*, but the easiest way to differentiate between the two is to do a spore print. *Cortinarius* has a rusty brown spore print, and purple-gilled *Laccaria* has a white to pale purple spore print. Many *Cortinarius* mushrooms are not edible, so always be sure!

RIGHT-QUICK RUNDOWN:

- light-purple to buff-white smooth cap
- purple, thick, distant gills
- long, scaly stem similar to color of cap
- growing scattered near hardwood trees

WRINKLED CORTINARIUS MUSHROOM
(CORTINARIUS CAPERATUS)

And since we are on the topic of underdog mushrooms, here's another: the wrinkled cortinarius mushroom. At least in my neck of the woods, I don't hear of many folks bringing these home. In other parts of the world, they are considered a delicacy. They have a very nice, mild mushroom flavor, makin' them a breeze to incorporate into a variety of dishes.

I've mentioned in previous mushroom profiles that the *Cortinarius* genus includes many mushrooms ya don't wanna muck with, but the wrinkled cortinarius mushroom is an exception. *Cortinarius* mushrooms often have spiderwebby-looking veils (called *cortinas*) that cover their gills when they are babies, and the veil later becomes a ring around the stem. The wrinkled cortinarius mushroom is atypical in that it is a member of the *Cortinarius* family but does not have the spiderweb remnants they are frequently known for. Instead, wrinkled cortinarius mushrooms have a very prominent annulus that to me can look just like someone slid a little ring onto the upper portion of the stem with no cobwebs detected.

There have been times in the fall I have headed out to go huntin' and won't see much fungi, but I seem to always see a few wrinkled cortinarius mushrooms. I have yacked with other foragers around me who say that wrinkled cortinarius mushrooms are rare for them. Maybe I just have the perfect settings or maybe they are drawn to me, but they've sure been faithful to me whatever the case. I especially see them in mixed forests that include beech and hemlock trees in moist areas around creeks or lakes.

Here, you can see that distinctive frosty white center that many wrinkled cortinarius caps will have.

Young specimens of wrinkled cortinarius mushrooms will start off having a rounded cap, later becoming flatter with age.

LOOKS:

They typically grow to have a cap up to 5 inches (13 cm) wide and stand up to 6 inches (15 cm) tall. The cap starts out rounded when young and flattens with age. The cap is colored pale yellow, turning orangish brown and exhibiting wrinkles when more mature. The cap often has a frosty white umbo in the center. It has gills that are white and covered by a veil when young and then the gills turn browner with age. The gills are attached to the stem and closely spaced. The stem is off-white is often uniform in shape. After the veil breaks away to reveal the gills, there will be a prominent ring on the upper portion of the stem.

STOMPING GROUNDS:

They prefer to grow in mixed forest containing both hardwood and conifer trees. I find many of mine in hemlock forests, but also around beech, birch, and oak. They have also been noted to grow around blueberry bushes.

GROWTH PATTERNS:

They can be found growing solitarily, scattered, or gregariously on the forest floor.

SPORE PRINT COLOR:

tawny brown

FLAVOR NOTES:

mild, nutty

POTENTIAL LOOKALIKES:

Some fieldcap mushrooms (*Agrocybe* sp., varying edibility throughout the genus) can resemble wrinkled cortinarius mushrooms. Fieldcaps often grow earlier in the year in woodchips or more open areas, unlike the wrinkled cortinarius mushroom. Field caps also have a distinctive smell of fresh ground flour or oatmeal.

The corrugated cortinarius (*Cortinarius corrugatus*, not edible) can also be a lookalike for the wrinkled cortinarius, but the main difference is the corrugated cortinarius lacks a ring around the stem.

RIGHT-QUICK RUNDOWN:

- yellowish to orange cap
- white frosty umbo
- closely spaced white to off-white gills
- prominent ring around the upper portion of stem

SHAGGY STALKED BOLETE
(AUSTROBOLETUS BETULA)

The shaggy stalked bolete is another one of those mushrooms that seemed surreal to me. It's so darn neat and intricate, and it looks like it's from another world. I remember the first time I ever saw a shaggy stalked bolete in real life, and the picture at left is *the* one! I was just a bee-boppin' down the trail and caught a glimpse of that lanky sucker stickin' up over the hill. I shouted, "SHAGGY STALKED BOLETE!" and scared the absolute crap out of my partner. I swiftly galloped down the hill like a Clydesdale and did a pretty impressive slide to stop myself that I will never be able to duplicate again in my life. I then got to appreciate that beautiful mushroom up close and snap a picture. It was *perfect*.

Shaggy stalked boletes make their presence known. They are looong boys. With their bright red and yellow coloration, they stick out like a sore thumb amongst the leaf litter in the fall. You'll see it wears a tiny little cap, but the mighty stem often allows it to tower over the forest floor.

With its wild, cartoonish appearance, it's only natural that it would have a kinda funky taste. The flavor is tart and lemony. The cap is soft and spongy in texture, but the stem stands tall and firm. When cooked, the stem is my favorite. It retains a lovely crunchy texture, sorta like sautéed asparagus. I find this mushroom pairs real nice with fish since it has that acidic flavor that goes so well with seafood.

Shaggy stalked boletes are so tall and colorful, makin' 'em hard to miss.

LOOKS:

They typically grow to have a cap width of up to 3 inches (7.5 cm) and stem length up to 8 inches (20 cm). The cap is rounded, reddish orange in color, and it often will have a bright yellow cap margin. The cap is spongy and soft to the touch. In wetter conditions, the cap appears moist and sticky. The underside is porous and yellow, becoming more greenish yellow with age. It does not bruise when cut or damaged. The stem is long, firm, and uniform in shape with deep reticulation. The reticulation appears as a rigid yellow overlay revealing a base color of red on the stem.

STOMPING GROUNDS:

They prefer to grow in mixed forests, especially amongst oak, beech, hemlock, and pine.

GROWTH PATTERNS:

They can be found growing solitarily or scattered on the forest floor.

SPORE PRINT COLOR:

olive green

FLAVOR NOTES:

tart, lemony, acidic

POTENTIAL LOOKALIKES:

Russell's bolete (*Aeroboletus russellii*) could be similar-ish to the shaggy stalked bolete. The main differences are that Russell's bolete is overall more reddish in color and has a distinctive scaly-appearing cap that gets wider in size. Shaggy stalked bolete exhibits more yellow coloring and has a smooth, smaller cap.

RIGHT-QUICK RUNDOWN:

- small reddish-orange cap with yellow margin
- yellow pores
- looong, deeply reticulated yellow and red stem
- does not bruise

CHAPTER 8

Edible Mushrooms of the Winter

Hot dang! Look at ya. You've made it to the final season, winter. As I mentioned before, I ain't never not foragin' and the wintertime is no exception. Many folks think because of some cold temperatures and snowfall, there's no mushrooms to be had, but buddy are they wrong. I *love* to mushroom hunt in the winter. It has many of my favorite edibles. There's just somethin' about layering up and hittin' the trails that warms my soul right up. Seasonal Affective Disorder (SAD) is real during this time of year, and my cure to offer ya is foraging. It sends my serotonin soaring through the forest canopies with every yummy mushroom I find. So, lace up them snow boots, throw on a boggin hat, and get ready because it's about to get tasty up in here.

Scan here to see some foraged goodies to snag in the cold weather!

LEFT | Don't ya worry. It's "snow" problem findin' edible mushrooms in the wintertime.

LATE FALL OYSTERS (SARCOMYXA SEROTINUS)

Yep! There are still more oysters to be had, honey! Meet the late fall oyster. Yeah, yeah, I know it's called the late "fall" oyster, but I have continued to find these puppies well into the winter year after year. They love to come out to play after the first frost.

They look a heck of a lot like other oyster mushrooms, but the main difference is their olive greenish to tannish caps. When you flip them over, they have those iconic decurrent oyster gills, but maybe a little more yellowish to orange than the whiter oyster gills you're used to. Don't let the green caps fool ya; they aren't rotten. Well, unless they smell and feel rotten, then they are probably rotten. A fresh specimen should be firm and dense in texture and have no distinctive odor. The caps often appear kinda slimy when wet. I think the caps resemble a wet toad. I know that doesn't sound appetizing but trust me here.

I have harvested many a late fall oyster that were frozen right to the wood. It's easy. You simply snap them off and throw 'em in your sack. Just use your judgement in seeing if the mushroom looks to be in good shape, and there's no harm nor foul in bringin' a healthy, frozen nugget home and letting it thaw. It's like nature preserved it just for you.

The flavor of late fall oysters is mild, so it's easy to toss into your dishes. One thing to note with this guy is he is kinda tough. His texture can be somewhat rubbery, so I like to cook them a little longer than my average wild mushroom to soften them up more.

Edibility has been debatable and all over the place with this one. There was some gossip goin' around that this mushroom shouldn't be eaten due to containing carcinogens, but that's just a rumor. You can safely consume this feller without fear.

Late fall oysters differ from your run-of-the-mill earlier season oysters in that they have greenish caps and more yellowish to orange gills.

LOOKS:

They typically grow as overlapping semicircular-shaped clusters, with each singular growing up to 5 inches (13 cm) wide. The cap can range in color from a deep olive green to a tannish brown. The cap is smooth and may appear sticky or slimy when wet. The underside reveals closely spaced yellowish to orange gills that are slightly decurrent if a stem is present. Late fall oysters often lack a stem, but if present, the stem is short, stubby, and can appear scaley in texture. The mushroom itself is thick, dense, and rubbery. When doing a cross section, the inner flesh is white.

STOMPING GROUNDS:

They prefer to grow on dead/dying hardwood trees, especially beech, oak, and poplar.

GROWTH PATTERNS:

They can be found growing solitarily, scattered, or in overlapping shelving clusters from hardwood.

SPORE PRINT COLOR:

white to yellowish

FLAVOR NOTES:

mild, mushroomy, more bitter with age

POTENTIAL LOOKALIKES:

Mock oysters (*Phyllotopsis nidulans*, not edible) could be a potential lookalike of the late fall oyster. The differences are that mock oysters are more orange in color and smell like trash.

RIGHT-QUICK RUNDOWN:

- olive green cap to tannish caps (sticky when wet)
- yellowish to orange closely spaced, decurrent gills
- lacking stem (or very small, stubby stem if present)
- growing from hardwood

CHAGA, MEDICINAL (*INONOTUS OBLIQUUS*)

Chaga is a highly coveted medicinal fungus. It's a little different than the others mentioned in this book. It doesn't present as your average toadstool, but instead as a big ol' hunk that grows often from wounds on the sides of trees, especially birch. We call this big ol' hunk a *conk*. It's basically a wad of blackened fungus that looks like a burnt tree burl. When you crack it open, it reveals a bright orange and corky interior.

Chaga can be spotted and harvested in other months, but winter seems to be the preferred time. When the foliage has fallen off the trees, they are easier to see. In colder northern regions, people like to go for hunts in the snow because the chaga stand out more during snow falls.

When harvesting chaga, you're probably gonna have to call in some backup. This stuff is *tough*. If I know I'm going to a birchy forest the time of year that chaga could be spotted, I will bring a hammer in my backpack. Yes, a hammer. Some folks even carry a small hatchet or a chisel of sorts to whack the stuff off the trees. A few good swings and you should be able to get some chunks off. Chaga can grow up high on trees and out of the reach of human hands. If you're gonna be climbin' a ladder or standin' on your buddy's shoulders to get this stuff, please be careful! I also like to leave some of the chaga behind (at least 2 to 3 inches [5 to 7.5 cm] of it) because chaga can regrow there, albeit slowly. It's totally cool to take chaga as it will not cause any further damage to the tree. If chaga is found on a dead, fallen tree, it is often too old. A good way to check is to see if the inside is black. If so, pass on it. If bright orange, you're good! I prefer to harvest from standing trees.

Once you get your chunks home, it's best to dry them immediately. I use canned air to blow off any dirt and debris. I don't wash it or get rid of any of the black exterior. That stuff is still gold! I do my best to break it into smaller pieces and dry it in my dehydrator on the lowest heat setting until it's done. I then store mine in mason jars, but you could use any airtight container you'd like. You then have dried chaga at the ready!

It boasts a ton of medicinal benefits and is best used as a tea or in a tincture or a powder form to receive those benefits. It has a rather bitter taste and a tough texture, so adding it to your ramen is gonna be a no-go here. Some benefits of chaga include boosting the immune system; lowering cholesterol, blood sugar, and blood pressure; slowing the growth of cancer cells; reducing inflammation; and combating anxiety, depression, and fatigue. As always, when it comes to using mushrooms medicinally, check to see if this mushroom will have any adverse reactions with prescribed medications or medical conditions.

Here, you can see the orangey innards of chaga.

LOOKS:

Chaga can come in all kinds of shapes and sizes, so to describe an average shape/size of growth is purdy impossible. It presents as a dark, rough growth on the side of the host tree, reminiscent of burnt charcoal. It is often lumpy and hard to the touch. The exterior is black and the interior of healthy chaga is a bright orange with a corky texture.

STOMPING GROUNDS:

They prefer to grow nearly exclusively from wounds on the trunks of birch trees. Chaga *can* be found growing on other hardwoods (elm, ash, beech, maple, alder, and poplar), but only birch tree chagas contain all the medicinal benefits previously mentioned.

GROWTH PATTERNS:

They can be found growing mostly solitarily on the host tree.

SPORE PRINT COLOR:

Chaga is non-spore producing, so no spore print color here!

FLAVOR NOTES:

mild, mushroomy, more bitter with age

POTENTIAL LOOKALIKES:

Sometimes regular ol' tree burls or knots can trick you into thinkin' you may have spotted chaga. Be mindful of what type of tree it is growing on to rule it out. Also check for the orange interior. Black knot fungus (*Apiosporina morbosa*, not edible) is another potential look alike, but it is commonly found on fruit trees and not birch trees.

RIGHT-QUICK RUNDOWN:

- black, bumpy growth (looks like charcoal)
- very hard texture
- growing on birch tree trunk
- orange, corky interior

ENOKI (*FLAMMULINA FILIFORMIS*)

This lovely cluster of flamin' charm is wild enoki. It's a highly sought after winter edible for many foragers. Some refer to them as velvet shank or velvet foot mushrooms due to their silky stems. They were previously referred to as *Flammulina velutipes* and *velutipes* means "with velvet legs," so it all makes sense. Whatever ya wanna call it, one thing that is universal is that it's so uplifting to see that explosion of sunshine in the hills on a cold, dreary day. Enoki can fruit well into the early spring, often before morels start to grow. Something else I've also noticed is that when I find a cluster of enoki, I also often find oyster mushrooms nearby or even on the exact same tree. Keep that head on a swivel!

If you have been to the grocery store or gone to a restaurant and gotten enoki mushrooms in your ramen, you may have noticed it looks quite a bit different than the wild enoki pictured here. It usually looks like long, thin, noodle-y, pale white mushrooms. There's a reason for that. These white enoki are cultivated indoors with high levels of carbon dioxide and they don't get to see the sunshine. These conditions cause them to grow in this manner and not develop those beautiful enoki colors you get in the wild.

Now, enokis do require some TLC. Given its sticky cap and tough stems, you'll have to do a few things before ya get to cookin'. First off, the stems are not always a pleasant texture, so many people toss them. If I find specimens that are young enough, I'll keep the stems. They lend a nice lil' crunch. If the stems are very tough and black, I will pass on them. Due to the caps being super viscid, they can be a pain to clean. Just take your time pickin' off dirty bits from the cap while holding it under running water. Another issue with the sticky cap is if you sauté without doing a par-dry, they will be pretty slimy. This will make your skillet looked like someone hocked a loogie in it. Definitely not appetizing. What I like to do is clean them off, stick them in a dehydrator for a bit, and *then* sauté them. If you don't have time for the par-dry, enoki is best dropped into soups, stews, and brothy noodle dishes to help mask that slimy texture.

This lil' cluster of enoki displays those beautiful caps that look like they got a fresh spit shine.

Scan here to look at a big, surprise enoki haul!

Peep this massive cluster of enoki fruitin' from its favorite, an elm tree.

LOOKS:

They typically grow in clusters on wood and can have a cap width up to 3 inches (7.5 cm) wide. The caps are yellowish orange in color, with the darkest of the color range more concentrated to the middle portion of the cap, lightening at the margins. The caps are more rounded when young, flattening out with age. The caps are often shiny and sticky. The gills are off-white to yellow, browning with age. The stem can grow to be up to 4 inches (10 cm) long and is paler yellowish orange in color when young, eventually turning dark brown to black when older, with very fine, velvety hairs. The lower portion of the stem closest to its attachment to wood is often slightly darker in color than the upper portion closest to the cap. The stem is hollow throughout when very mature.

STOMPING GROUNDS:

Enoki prefer to grow on elm trees especially but can grow from persimmon and mulberry trees, too. They also like to grow behind shedding bark of these trees at times, so I always check behind the bark when I can.

GROWTH PATTERNS:

They can be found growing sometimes gregariously, but most often as clusters from hardwood.

SPORE PRINT COLOR:

white

FLAVOR NOTES:

mild, earthy, slightly sweet

POTENTIAL LOOKALIKES:

Galerina marginata, also called deadly galerina or the funeral bell (if that gives you any indication of whether you should eat it or not) is a toxic lookalike of enoki. Don't worry though, with the 411, they are easy to tell apart. The main difference is the spore print. Enokis throw a white spore print, whereas deadly galerina throws a rusty brown spore print. Some other differences include that deadly galerina have a ring around the stem, the stem often appears shaggy below the ring, and they have darker, more helmet-shaped caps that aren't slimy. A quick supplemental trick I like to recommend if you are tryin' to figure these two out: does it stick to your face? What the heck am I talking about? I mean, pluck a mushroom and stick the cap to your cheek. Does it stick? Probably an enoki. Does it not stick? Probably deadly galerina.

RIGHT-QUICK RUNDOWN:

- yellowish-orange shiny caps
- growing in clusters from hardwood
- dark stems (no ring) with fine hair
- off-white to yellow gills

Edible Mushrooms of the Winter

YELLOWFOOT (*CRATERELLUS TUBAEFORMIS*)

I'll tell ya what, I love me some yellowfoot. To be such a small thang, it packs a hefty flavor. You may also hear them called *winter chanterelles*, but both common names are fitting. It gets the name "yellowfoot" from its yellow stem. It gets the name "winter chanterelle" because, well, it's a chanterelle that grows in the wintertime. I have found yellowfoot in the late fall and early spring as well. I'll take 'em whenever I can get 'em.

Funny story about yellowfoot right-quick. When I first started learnin' about winter mushroom foraging, I had been studying these yellowfeet hard one day because I wanted to get my hands on some. That night, I went to bed, and I dreamed all night long that they were all over a certain trail I hike that had all the right trees and terrain. The next mornin' when I woke up, I told my partner immediately that my shroomy senses were tingling and we needed to go to that trail and check for yellowfoot. We geared up and away we went. Guess what I found when we got there? *A ton of yellowfoot*! I got to fill up my sack with 'em and cook 'em, and I've been hooked ever since. I guess I'm some sort of mushroom clairvoyant. Who knew?

These mushrooms tend to grow in gobs, which is another thing I love about them. They are small and delicate, so you need yourself a good haul of them to make a dish worthwhile. Mother Nature comes through and tends to scatter these all over the hills for me come November and December.

I have my best luck finding yellowfoot in coniferous forests, heavy on the hemlock and pine. Bonus points if you can spot mountain laurel and moss as well. Yellowfoot love a downed, mossy conifer tree. If I see a pine tree that has fallen and started to decay, that's the stuff right there. I will check on the ground all around the length of the fallen tree. Don't forget to check both sides. These mushrooms can hide on ya, so a little shuffle of the leaf litter right by the dead tree can reveal a clump of yellowfoot. I liken the experience to unwrapping a precious gift. It makes me giddy as all get out when I move some leaves and *kerpow*! A cluster of yellowfoot.

Contrary to popular belief, chanterelles *can* grow on wood. I have found yellowfoot sprouting straight from fallen, dead conifers plenty of times.

Scan here to check out some yellowfoot mushrooms and what I make with 'em!

I treat these a lot like I treat my black trumpets. They are easy to clean by snippin' off their dirty butts and givin' 'em a swish in cool water. They dry well, too. I love to use yellowfoot in rich, creamy, herby dishes. The smoky, peppery flavor pairs so well with many things. They are tasty enough that a simple sauté in some butter, garlic, salt, and pepper is even hard to beat. Lay those toasty mushrooms over a steak and shew-wee! It's so dang good.

WILD YELLOWFOOT QUICHE

INGREDIENTS
- 1 frozen 9-inch (23 cm) deep-dish pie crust
- ⅓ cup (53 g) thinly sliced shallots
- 4 cups (216 g) cleaned yellowfoot mushrooms
- 2 tablespoons (28 ml) olive oil
- Salt and pepper to taste
- ½ cup (120 ml) heavy cream
- ½ cup (120 ml) milk
- 3 large farm-fresh eggs
- Dash nutmeg
- 1 cup (120 g) grated gruyere cheese and 1 cup (120 g) smokey white cheddar cheese, mixed (or use whatever white cheese you love)

DIRECTIONS

1. Preheat the oven to 350°F (180°C, or gas mark 8). Place aluminum foil loosely over the top of a frozen pie crust and bake for 15 minutes.
2. Remove from the oven, let cool, and remove the aluminum foil. Lightly poke holes in the bottom of crust with fork and bake again at 350°F (180°C, or gas mark 8) until lightly golden.
3. Remove from the oven and set aside.
4. Sauté the shallots and mushrooms in olive oil in a skillet until the mushrooms begin to brown on their edges. Season with salt and pepper. Remove the skillet from the heat.
5. In a large bowl, whisk together the cream, milk, eggs, nutmeg, and salt and pepper to taste.
6. Get your prebaked pie crust and place it on a baking pan to prevent any overflow in the oven while baking.
7. Sprinkle half of the cheese on the bottom of the crust.
8. Spread the mushroom and shallot mixture evenly over the layer of cheese.
9. Add the remaining cup (120 g) of cheese on top of the mushroom layer.
10. Pour the egg and cream mixture over the layers (it will fit, I promise!).
11. Bake at 350°F (180°C, or gas mark 8) for 35 to 40 minutes or until set and the middle of the quiche is slightly wiggly.
12. Let cool for 20 minutes. (I know it's hard.)
13. Slice and serve!

Yellowfoot will often have a belly button in the middle of their cap connecting to the stem that is hollow throughout.

You can really see here how the mushroom gets its name. Note the yellow stem and iconic false gills.

LOOKS:

They typically grow to have small trumpet or vase-shaped caps that can get up to 3 inches (7.5 cm) wide. The cap surface is often brown and becomes grayer with age. The cap is smooth with wavy, thin margins. The center frequently has a depressed hole in the middle that connects to the hollow stem. The underside reveals distinct, false, forked, decurrent gills. The gill surface appears lighter in color than the cap surface, usually a grayish color, but sometimes browner depending on age. The stem is typically up to 3 inches (7.5 cm) long, is hollow throughout, mostly uniform in shape, and is yellow to orange in color. The flesh is very thin.

STOMPING GROUNDS:

They prefer to grow on or near rotted, fallen conifer trees, especially pine and hemlock. They also love moss.

GROWTH PATTERNS:

They can be found growing solitarily, most often scattered, or in loose clusters on the ground or fruiting directly from dead conifer wood.

SPORE PRINT COLOR:

white

FLAVOR NOTES:

earthy, smoky, slightly fruity and peppery

POTENTIAL LOOKALIKES:

Pinelitter gingertails (*Xeromphalina caucticinalis*, too bitter to eat) are a common lookalike of yellowfoot. Not only do they look similar, but they also grow in the same spots at the same time of year near each other. The differences are that pinelitter gingertails have a more orangey brown cap, a tougher brown stem that is often hairy-ish toward the bottom portion, and it will lack the center hole in the cap that yellowfoot has. Pinelitter gingertails are also not hollow throughout like yellowfoots.

RIGHT-QUICK RUNDOWN:

- yellowish-brown to dark-brown trumpeted cap
- belly button in the center of the cap
- thin
- forked, false gills
- yellow stem
- hollow throughout

HEDGEHOG MUSHROOM (HYDNUM SP.)

Okay, I am absolutely not one to discriminate against any wild, edible mushroom, but I'm gonna go ahead and say it: The hedgehog mushroom is *my* favorite mushroom. Yeah, it's delicious and cool lookin', but it's my favorite for another reason. During the time of year when you may start to experience the cold weather blues and might be missing your summertime mushrooms, the hedgehog mushroom comes to the rescue. It's an awful lot like the chanterelle mushroom in some of its features and in its flavor/texture profile. Studies have shown that the hedgehog may be related more so with chanterelles than other tooth fungi! I have found these puppies in freezing temperatures before and dusted snow off their caps to harvest them. It provides the best mood boost right when ya need it.

It is also probably one of the easiest beginner mushrooms out there. If it has an orangey top, a stem, and teeth/spines underneath, it's a hedgehog, buddy! There are no lookalikes to worry about with this one. To nail down an *exact* species of this mushroom is a little complicated. There's been some back-and-forth between *Hydnum repandum*, *Hydnum rufescens*, and *Hydnum umblicatum*, so we will refer to this hedgehog pictured and discussed as *Hydnum* spp. until further research indicates what we shall call this specific mushroom. Whichever it ends up being, they are all good, safe edibles.

The flavor of a hedgehog mushroom is outstanding! It is a lot like a chanterelle in that it is nutty and slightly sweet, and the texture is nice and sturdy, holding up well to all kinds of cookery. Not only does it taste like a chanterelle, but it can resemble chanterelles, too. In the fall and winter, you might see hedgehogs peeping through the leaf litter, and you'll swear it looks like chanterelle caps stickin' up. Flip 'em over and you'll find the surprise of a toothy spore-producing surface, which confirms it's a juicy hedgehog.

Something I have come to find in my years of trekking for these mushrooms is that they love to grow in a line. If I see one hedgehog, I will often see a line of them growing out in some direction from the original one I spotted. I call it "followin' the orange brick road." I don't have any science or research to put behind this observation, just lettin' ya know that mama sees it a lot, and it helps me to find more.

We talked about yellowfoot in the previous profile. Something to note is that I find hedgehog and yellowfoot mushrooms growing near one another every year. I do this so often that I'll usually bring two mushroom sacks: one for yellowfoot and one for hedgehogs! I look for hemlocky, mossy, moist forests with lots of downed conifers. It is a primo breedin' ground for these two tasty treats.

Don't let the toothy underside scare ya when it comes to cleaning your hoard. I like to use the sprayer on my sink to blast out any dirt from the underside. It works like a charm. And of course, I whip out my handy dandy mushroom toothbrush to scrub-a-dub the caps and stems as needed under running water. I've heard many people say that once the teeth on the mushroom have started to turn more orange, they become bitter. Folks recommend taking a spoon or knife and scraping off the teeth to rid it of the bitterness if this is the case. I'm just here to tell you that I've harvested hedgehogs with more orangey teeth before and I didn't find the taste to be off at all, so I always leave the teeth intact. If the teeth are obviously very brown in the field, I don't harvest them in the first place. Use your judgement.

HEDGEHOG MUSHROOM BACON PENNE PASTA

INGREDIENTS

- 1 box (16 ounces, or 455 g) penne pasta
- 6 pieces of thick sliced bacon
- 1 small onion, diced
- 6 cloves minced garlic
- 4 cups (216 g) cleaned and chopped hedgehog mushrooms
- 1 cup (235 ml) pasta cooking water
- 1 cup (235 ml) heavy cream
- 2 sprigs fresh thyme
- 2 tablespoons (8 g) chopped fresh oregano
- 1 tablespoon (14 g) butter
- Handful fresh chopped parsley
- ½ cup (40 g) shredded Parmesan

DIRECTIONS

1. Cook the penne pasta according to the instructions on the box. Strain, saving 1 cup (235 ml) of the pasta cooking water for later, and set aside.
2. Dice the bacon and cook in a skillet on medium-high heat until crispy. Set aside. Save the bacon grease in the skillet (duh!).
3. Add the onion and garlic to the skillet with the bacon grease and cook for 2 minutes, stirring occasionally.
4. Add the hedgehog mushrooms and reduce the heat to medium. Sauté for 10 minutes.
5. To the skillet, add the reserved pasta water, heavy cream, thyme, and oregano. Bring to a low boil.
6. Once boiling, add the penne pasta. Stir and let cook until the sauce begins to thicken.
7. Add the butter, parsley, Parmesan cheese, and chopped bacon and stir until well incorporated.
8. Serve it up!

The toothy underside of the hedgehog makes it pert near impossible to misidentify. Those lil' quills under the cap are its trademark.

LOOKS:

They typically grow to have a cap width of up to 4 inches (10 cm) and stand up to 5 inches (13 cm) tall. The cap is smooth and very pale orange to darker orange in color, depending on age. The cap margins become wavier when mature. The cap often has a little hole or indent in the middle, like a belly button. The underside has many teeth, or quills (hence the name hedgehog). When younger, the underside is cream-colored, and with age, the teeth and spore-producing surface may turn more orangish brown. The stem is off-white to pale orange in color, smooth, and thick in texture.

STOMPING GROUNDS:

They prefer to grow around conifers, especially hemlock and pine, and can also fruit around spruce, birch, oak, and beech trees. They love moss.

GROWTH PATTERNS:

They can be found growing solitarily, often scattered (in lines) to gregariously from the forest floor.

This is a common way to find hedgehogs growin', scattered amongst the leaf litter.

SPORE PRINT COLOR:

white

FLAVOR NOTES:

slightly sweet, nutty (a lot like chanterelles)

POTENTIAL LOOKALIKES:

Ain't none!

RIGHT-QUICK RUNDOWN:

- pale-orange to orange smooth cap
- similarly colored stem
- many small teeth or "quills" under the cap

TURKEY TAIL, MEDICINAL
(TRAMETES VERSICOLOR)

If someone asked me what mushroom is the easiest to find, I'd probably holler, "Turkey tail!" This stuff grows *everywhere*. Basically, if there's wood, you bet your hind end there's gonna be some turkey tail. I'd put a lot of money on me finding it any time of year anywhere in the woods. Though it can be spotted in all seasons, I have placed it here in winter because this is when I tend to find the nicest, freshest specimens.

Turkey tail appears as beautiful little fan-shaped overlapping shelves or as bloomin' rosettes on wood. The color variation is insane with these. They have those zonations we talked about to the nth degree. They show off strips of rotating colors ranging anywhere between brown, red, and blue, and they also have alternating textures of furry smoothness.

Turkey tail is known for its medicinal benefits, so it is best utilized in extractions. Its popularity has really exploded over the last few years. People seek it out to use for immunity, gut health, cancer, mood improvement, and for its antibacterial and antiviral properties.

Turkey tail is very tough and rubbery in texture, so adding it to your supper plates ain't really the best place for it. However, I have tossed some dried turkey tail into veggie stocks and bone broths for an extra added healthy benefit before. Otherwise, you'll catch me tincturin' mine most of the time. *Also,* just so ya don't make the same mistake I did, I wanted to share something. I dry and powder a lot of medicinal plants and mushrooms, but let me tell ya, turkey tail ain't the one. I found it out the hard way. When you dry turkey tail and put it in a spice grinder, it turns into a giant cotton ball. You have been warned.

When it comes to harvesting turkey tail, you're liable to find it in all kinds of different maturation stages. My go-to way of gauging if I want to take it home is to take a look under the hood. If you pluck off a piece of turkey tail, a choice, fresh specimen will have a snow white, porous underside. If the underside has started to become yellow or brown, it's past its prime, so I leave it be.

Turkey tail has a wide range of colors to show off. Here is a more neutral lookin' cluster of brown and white turkey.

LOOKS:

They typically grow as overlapping shelves or rosettes with each cap growing up to 4 inches (10 cm) wide. The surface of the cap is distinctly zonated with varying colors and textures, ranging from tans to browns to reds to blues to white to greens, often with fine hairs, giving it a velvety feel. The outer margin of the cap is usually lighter in color. The underside is white when fresh, turning yellow or brown with age, with teeny pores. It has no stem and attaches broadly to wood. The texture of the mushroom is rubbery and leathery, being easy to bend.

STOMPING GROUNDS:

They prefer to grow on fallen logs, tree stumps of hardwood, and even on living hardwood trees, occurring less frequently on conifers.

GROWTH PATTERNS:

They can be found growing in overlapping shelves or rosette clusters from wood.

SPORE PRINT COLOR:

white

FLAVOR NOTES:

earthy, woody (too chewy to eat, so it is often used medicinally in teas/tinctures)

POTENTIAL LOOKALIKES:

False turkey tail (*Stereum ostrea*, not edible) is a common lookalike of real-deal turkey tail. The main differences are that false turkey tail is thinner and more brittle in texture, does not have the range of colors as true turkey tail (usually gray to reddish browns), does not have pores on the underside, and the underside is smooth, which will often allow you to see faint zonations from the top of the cap peeking through the underside.

RIGHT-QUICK RUNDOWN:

- zonated, fan-shaped cap of alternating colors and textures
- velvety feel
- white underside with small pores
- growing from wood in shelves or rosettes

Edible Mushrooms of the Winter

DOUBLE-EXTRACTION TURKEY TAIL TINCTURE

INGREDIENTS

- Dried or fresh turkey tail mushroom (cut into small pieces)
- 80 proof or higher vodka (You can use Everclear or moonshine)
- Filtered water

DIRECTIONS

1. Cut the fresh or dried turkey tail mushrooms into smaller pieces with sharp scissors (the smaller the pieces, the more surface area for a better medicinal extraction).
2. Fill a mason jar (whatever size you like) either halfway full with dried turkey tail mushrooms or three-quarters of the way full with fresh turkey tail.
3. Pour in the alcohol to reach the top of the jar and completely cover the turkey tail.
4. Cover with a lid and store in cool, dark place for 4 to 6 weeks, shaking daily.
5. After 4 to 6 weeks, pour the mixture through a fine mesh strainer, reserving the turkey tail mushrooms and setting to the side.
6. Pour the strained alcohol liquid into your jar and save.
7. Get a large pot and add in your reserved turkey tail mushrooms and add double the amount of water that you have of the strained alcohol (e.g., If you have 1 cup [235 ml] of strained alcohol from the first extraction, add 2 cups [475 ml] of filtered water to the pot with reserved turkey tail).
8. Simmer the water and turkey tail for hours until the volume of water has reduced to half of your volume of the first alcohol extraction. For example, if you have 1 cup (235 ml) of alcohol extraction, you'll have ½ cup (120 ml) of the water extraction.
9. Strain out the turkey tail from water and let cool. Compost the turkey tail.
10. Once the water extraction has cooled, mix with the alcohol extraction. You now have a double extraction!
11. Bottle the tincture in amber glass dropper bottles. Use as needed.

Note: A typical dose of turkey tail tincture is 15 to 30 drops under the tongue or diluted in a beverage up to 2 times daily as needed.

WITCHES' BUTTER (TREMELLA MESENTERICA)

This big ball of yellow snot is witches' butter. It's another type of edible jelly fungi. Sorry I called it snot, but I'm a straight shooter. *pew pew*

There are quite a few mushrooms that may also be called witches' butter, but this is *the* witches' butter to me and most.

In the winter, it will be easy to spot in the snow as a golden, frozen glob on sticks and branches of hardwood trees.

Witches' butter is available year-round, just like turkey tail, but witches' butter is a bit tougher. It can withstand multiple dry spells, freezes, and the like. It will dry up and reconstitute like nothin' ever happened. I admire the fortitude of witches' butter.

Since it's an edible jelly mushroom, it's best added to soups, stews, and broths for texture, and like amber jelly roll, it can be turned into gummy candies. Its flavor is bland, but it's a good survival mushroom if need be.

Witches' butter has also been discussed as a medicinal mushroom that has anti-inflammatory properties and is used for circulation, skin health, and respiratory health.

You'll often find witches' butter growing in their own separate blobs along tree branches and sticks.

LOOKS:

It typically grows as a gummy glob of lobey folds on wood, up to 4 inches (10 cm) across. It is golden yellow to orange in color, smooth, and rubbery in texture. When fresh, it will be wet and moist. In dry conditions, it will be thin and brittle, and in freezing temperatures, it will be frozen of course.

STOMPING GROUNDS:

They prefer to grow mostly on fallen sticks and branches of hardwood trees, sometimes on fallen logs.

GROWTH PATTERNS:

They can be found growing solitarily or scattered from hardwood.

SPORE PRINT COLOR:

yellowish

FLAVOR NOTES:

doesn't have a ton of flavor and takes on the flavor of whatever sauce/broth/seasoning you cook it in

POTENTIAL LOOKALIKES:

Orange jelly fungus (*Dacrymeces palmatus*, edible) is a common lookalike of witches' butter. The main difference here is color. Witches' butter is more golden yellow in color, while orange jelly fungus is, well, orange. When witches' butter is found dried out in the forest, its color can appear more orangey, but when rehydrated, it will be yellow.

RIGHT-QUICK RUNDOWN:

- globular, brainy, yellow- to golden-colored structure
- rubbery texture
- moist in wet conditions or dehydrated in dryer weather conditions
- growing on hardwood sticks/branches and sometimes fallen logs

CONCLUSION

Phew-wee! I'm awful windy when it comes to foragin', but would ya looky there?! Ya done made it, honey! How ya feelin'? Ya ready to bust them hills wide open?

If you didn't already know it before you cracked this book, ya now know there are sooo many wild and edible mushrooms out there just waitin' on ya. Hopefully, you've picked up a trick or tip or two as your eyes and brain traveled through the previous pages. I hope my words have helped you and eased your mind while also getting ya pumped up to do the dang thing. You totally *can* do it. Remember, the fungi kingdom wasn't built in a day, so ya got plenty of time for this. Before you know it, you'll be identifying and cookin' up mushrooms like a boss!

So, whenever the spirit moves you and you're feelin' right froggy, just go on and jump. Put all your learnin', research, experience, and knowledge to work and . . .

GO FORTH AND FORAGE!

Resources

Mushrooms of the Southeast
by Todd F. Elliot and Steven L. Stephenson

Appalachian Mushrooms
by Walter E. Sturgeon

National Audubon Society Field Guide to North American Mushrooms
by National Audubon Society

About the Author

Appalachian Forager Whitney Johnson is a full-time forager, content creator, and Appalachian ambassador. With over 1.5 million followers across her social media, she has used her platform to shine a lovin' light on Appalachia and to teach the world how to be more self-sufficient while responsibly utilizing the goodies that Momma Nature provides.

Born and raised in eastern Kentucky, she developed a passion for all things outdoors as a young holler baby. Foraging is her forte, especially mushroom hunting, but she does not discriminate against any wild edibles. She also picks plants and herbs to wildcraft a plethora of skincare and wellness products. She fishes. She hikes. She gardens. She can make jam or jelly out of just about anything. She cooks like your mamaw. She is a proud Appalachian spreading the love and wondrous things the region has to offer via her down-to-earth, educational, and quirky videos on TikTok (@appalachian_forager), Instagram (@appalachian_forager), and Facebook (Appalachian Forager).

Her accolades include being named Appalachian Arts and Entertainments Awards' Best Social Media Influencer, Appodlachia's Appalachian of the Year and Content Creator of the Year, as well as being featured in numerous news segments and articles, including multiple appearances on The Weather Channel. Whitney continues to be a hillbilly force to be reckoned with, spreadin' pride for her region, preachin' girl power, all while keepin' the old ways alive—with a dash of humor.

Index

A

adnate gills, 29, 144
adnexed gills, 29
Agaricus, 65
Amanita
    destroying angel (*Amanita bisporigera*), 41
    eastern Caesar's amanita (*Amanita jacksonii*), 38
    fly agaric (*Amanita muscaria*), 25
    grisette (*Amanita vaginata*), 26
    as puffball lookalike, 94, 97, 98
    as shrimp of the woods lookalike, 146
    universal veil, 37
amber jelly roll (*Exidia recisa*), 36, 67, 148–149
Amber Jelly Roll Gummy Candies, 148
American yellow fly agaric (*Amanita muscaria* var. *guessowii*), 25
anatomy
    annulus, 37
    basal bulbs, 38
    caps, 25–27, 33
    diagram, 24
    gills, 16
    hyphae, 39
    mycelium, 12, 38, 39, 43
    pores, 33
    reticulum, 38
    rings, 37
    scales, 26, 38
    skirts, 37
    spores, 28–31, 34
    stalks, 37
    stems, 37, 38
    stipes, 37
    striations, 26
    teeth, 33
    volvas, 38
    warts, 25
    zonations, 27
angel wing (*Plueocybella porrigens*), 116
animals, 20–21
annulus, 37
Appalachian chanterelle (*Cantharellus appalachiensis*), 78–79
Appalachian truffles (*Tuber canaliculatum*), 34
Ascomycotas, 34
attached gills, 29

B

basal bulbs, 38
Basidiomycota family, 35, 36
bear Lentinus (*Lentinellus ursinus*), 129
bear's head tooth (*Hericium americanum*), 33, 135, 136–137
beefsteak polypore (*Fistulina americana*), 33, 100–101
Berkeley Jerky, 110
Berkeley's polypore (*Bondarzewia berkeleyi*), 89, 110–111
biting insects, 20
black knot fungus (*Apiosporina morbosa*), 169
black-staining polypore (*Meripilus sumstinei*), 111, 132
black trumpets (*Craterellus fallax*), 82–84, 117, 130
blue chanterelles (*Polyozellus multiplex*), 83
boletes
    butter-foot bolete (*Boletus auripes*), 120
    porcini (*Boletus edulis*), 120
    Boletus flammans, 122
    Boletus rubroflammeus, 122
    Frost's bolete (*Exsudoporus frostii*), 120, 121–122
    old man of the woods (*Streobilomyces floccopus*), 102–103
    ornate-stalked bolete (*Retiboletus ornatipes*), 120
    painted bolete (*Suillus spraguei*), 120
    pores, 33
    reticulated stems, 38
    Russell's bolete (*Aeroboletus russellii*), 163
    shaggy-stalked bolete (*Aureoboletus betula*), 38, 120, 162–163
    trees and, 12
brain puffball (*Calvatia craniiformis*), 98
bruising, 31
bulbous honey fungi (*Armillaria gallica*), 37, 143–144, 145
butter-foot bolete (*Boletus auripes*), 120

C

caps, 25–27, 33
cauliflower mushroom (*Sparassis americana*), 14, 152–153
chaga (*Inonotus obliquus*), 168–169
chanterelles
    Appalachian chanterelle (*Cantharellus appalachiensis*), 78–79
    black trumpets (*Craterellus fallax*), 82–84, 117, 130
    blue chanterelles (*Polyozellus multiplex*), 83
    Chanterelle Cream Sauce Linguine, 76

cinnabar chanterelle
  (*Cantharellus cinnabarinus*),
  80–81
false gills, 30
golden chanterelle
  (*Cantharellus* sp.), 8, 73–74,
  130
peach chanterelle
  (*Cantharellus persicinus*), 77
smooth chanterelle
  (*Cantharellus lateritius*), 77
trees and, 12
Wild Summer Mushroom
  Toast, 117
chestnut bolete (*Gyroporus smithii*), 120
chew and spit test, 42
chicken of the woods
  Berkeley Jerky variation, 111
  Chicken of the Wood All-
    Purpose Seasoning Salt, 86
  Latin binomial, 10
  pores, 33
  white (*Laetiporus cincinnatus*),
    10, 44, 88–89
  Wild Summer Mushroom
    Toast, 117
  yellow (*Laetiporus sulphureus*),
    10, 47, 86–87
cinnabar chanterelle (*Cantharellus cinnabarinus*), 74, 80–81
close gills, 32
clothing, 17, 20
club fungi, 16, 35, 36
common ink cap (*Coprinopsis atramentaria*), 105
common morel (*Morchella americana*), 6, 58, 60, 61
common names, 10
common puffball (*Lycoperdon perlatum*), 35, 92, 96–97
common stinkhorn, 35
community, 11

conifer trees, 12, 13
Coprinopsis atramentaria
  (common ink cap), 105
Coprinus comatus (shaggy mane),
  104–105
corals
  Basidiomycota family, 35
  as bear's head tooth lookalike,
    137
  coral tooth fungus (*Hericium coralloides*), 138–139
  crown-tipped coral (*Artomyces pyxidatus*), 123–124
  fire coral fungus (*Trichoderma cornu-damae*), 42
  as lion's mane lookalike, 135
  Ramaria genus, 76, 123
  strict-branch coral (*Ramaria stricta*), 124
  teeth, 33
  varieties of, 36
  yellow-tipped coral fungus
    (*Ramaria formosa*), 36
corrugated milk cap (*Lacatarius corrugis*), 31
Cortinarius
  gypsy mushroom (*Cortinarius caperatus*), 160–161
  as indigo milk cap lookalike,
    113
  as purple-gilled *Laccaria* lookalike, 159
  as wood blewit lookalike, 141
  wrinkled cort (*Cortinarius corrugatus*), 161
Cream of Wild Mushroom Soup,
  130
crotch rocket (*Lactifluus volemus*),
  90–91
crowded gills, 32
crown-tipped coral (*Artomyces pyxidatus*), 123–124
cup fungi, 34

D

deadly galerina (*Galerina marginata*), 16, 171
dead man's fingers (*Xylaria polymorpha*), 36
deciduous trees, 12, 13
decurrent gills, 28, 29
deer mushroom (*Pluteus cervinus*), 15, 29, 68–69
destroying angel (*Amanita bisporigera*), 41
devil's urn (*Urnula craterium*),
  34, 83
discouragement, 48, 50, 55
distant gills, 32
Double Extraction Turkey Tail
  Tincture, 180

E

earthballs (*Scleroderma citrinum*),
  94, 97, 98
eastern half-free morel (*Morchella punctipes*), 60–61
enoki (*Flammulina filiformis*), 11,
  16, 43, 170–171
ethical foraging, 21, 48
evergreen trees, 12, 13

F

fall mushrooms
  amber jelly roll (*Exidia recisa*),
    36, 67, 148–149
  cauliflower mushroom
    (*Sparassis americana*), 14,
    152–153
  coral tooth fungus (*Hericium coralloides*), 138–139
  gypsy mushroom (*Cortinarius caperatus*), 160–161
  hen of the woods (*Grifola frondosa*), 111, 131–133

lobster mushroom *(Hypomyces lactifluorum)*, 154–155
oysters *(Pleurotus ostreatus)*, 128–129
purple-gilled *laccaria (Laccaria ochropurpurea)*, 158–159
resinous polypore *(Ischnoderma resinosum)*, 156–157
shaggy stalked bolete *(Austroboletus betula)*, 38, 120, 162–163
shrimp of the woods *(Entoloma abortivum)*, 145–147
snow fungus *(Tremella fuciformis)*, 36, 150–151
wood blewit *(Collybia nuda)*, 113, 140–141, 142
false chanterelle *(Hydrophoropsis aurantiaca)*, 75, 76, 79
false gills, 30, 72
false turkey tail *(Stereum ostrea)*, 179
fangers, 36
fieldcap *(Agrocybe* sp.*)*, 161
field guides, 9–10
fire coral fungus *(Trichoderma cornu-damae)*, 42
fly agaric *(Amanita muscaria)*, 25
footwear, 17
foraging. *See also* harvesting
   animals, 20–21
   biting insects, 20
   clothing, 17, 20
   ethics, 21
   field guides, 9–10, 17–18
   footwear, 17
   home research, 14–15
   journaling, 14
   knives, 18, 44, 49
   littering, 21, 48
   locations, 14, 16–17
   mesh bags, 18, 49
   mosquitos, 20
   opportunities, 21
   pack, 17–18
   panoramic hunting, 13
   patience and, 13
   permission, 16–17
   plants to avoid, 21
   rules, 16
   simplicity, 49
   size variations, 53
   snakes, 19–20
   socks, 17
   spiders, 20
   ticks, 20
   trash collection, 19, 21
   trees, 12–13, 14
   trim and drop, 21
forked gills, 29, 36
free gills, 28–29
Frost's bolete *(Exsudoporus frostii)*, 120, 121–122
fun, 50
funeral bell *(Galerina marginata)*, 16, 171

G

Ganoderma sp.
   as beefsteak polypore lookalike, 101
   Gandoderma curtisii, 108
   Ganoderma lucidum, 108
   Ganoderma sessile, 108
   hemlock reishi *(Ganoderma tsugae)*, 107–108
   as resinous polypore lookalike, 157
giant puffball *(Calvatia gigantea)*, 93–95
gills
   adnate gills, 29, 144
   adnexed gills, 29
   attached gills, 29
   bruising, 31
   close gills, 32
   crowded gills, 32
   decurrent gills, 28, 29
   distant gills, 32
   false gills, 30, 72
   forked gills, 29, 36
   free gills, 28–29
   myths, 43
   spacing, 32
   subdistant gills, 32
   toxicity and, 43
   true gills, 30
   universal veil, 37
golden chanterelle *(Cantharellus* sp.*)*, 8, 73–74, 130
Grilled Puffball Steaks, 95
grisette *(Amanita vaginata)*, 26
gypsy mushroom *(Cortinarius caperatus)*, 160–161
Gyromitra, 58

H

half-free morel *(Morchella punctipes)*, 57, 60–61
hardwood trees, 12, 13
harvesting. *See also* foraging
   cutting vs. plucking, 44
   mushroom population and, 43
   mycelium and, 39, 43
   overexploitation, 48
   pick shaming, 44
   regeneration, 44
Hedgehog Mushroom Bacon Penne Pasta, 176
hedgehog mushroom *(Hydnum* sp.*)*, 12, 33, 175–177
hemlock reishi (Ganoderma tsugae), 107–109
hemlock trees, 12
Hen of the Wood Flatbread, 133
hen of the woods *(Grifola frondosa)*, 111, 131–133
hunting. *See also* foraging
Hygrocybe, 81
Hygrophorus, 81
hyphae, 39

I

identification
   chew and spit test, 42
   confidence and, 45, 50
   gills and, 28
   home study, 14–15, 41
   identification (ID) groups, 11
   questionable mushrooms, 41
   spore prints, 15–16
indigo milk cap *(Lactarius indigo)*, 31, 112–113. *See also* milk cap *(Lactifluus volemus)*

J

jack-o'-lantern mushroom *(Omphalotus illudens),* 30, 75
jelly fungi
    amber jelly roll *(Excidia recisa),* 36, 67, 148–149
    Basidiomycota family, 35
    identification, 36
    orange jelly fungus *(Dacrymeces palmatus),* 182
    snow fungus *(Tremella fuciformis),* 36, 150–151
    witches' butter *(Tremella mesenterica),* 181–182
    wood ear *(Auricularia auricula-judae),* 36, 66–67
journaling, 14

K

knives, 18, 44, 49

L

late fall oysters *(Sarcomyxa serotinus),* 166–167. *See also* oyster mushrooms *(Pleurotus genus)*
latex, 31, 91, 112
Latin binomials, 10, 136
Lion's Mane "Crab" Cakes, 133
lion's mane *(Hericium erinaceus),* 33, 51, 133, 134–135, 137, 139
littering, 21, 48
Lobster Mushroom Duxelles, 154–155
lobster mushroom *(Hypomyces lactifluorum),* 154–155

M

mayapple *(Podophyllum peltatum),* 56
mesh bags, 18, 49
milk cap *(Lactifluus volemus),* 12, 31, 43, 90–91. *See also* indigo milk cap *(Lactarius indigo)*
mock oyster *(Phyllotopsis nidulans),* 129, 167
morels *(Morchella sp.)*
    Ascomycota family, 34
    categorization, 28, 34
    common morel *(Morchella americana),* 6, 10, 57, 58, 60, 61
    Cream of Wild Mushroom Soup, 130
    eastern half-free morel *(Morchella punctipes),* 60–61
    Fried Morels, 59
    lookalikes, 58, 60, 61
    name variations, 10
    raw consumption of, 47
    soaking, 59
    spore prints, 16
    toxicity of, 47
    yellow morel *(Morchella americana),* 57
mosquitos, 20
mushroom identification (ID) groups, 11
mushroom knives, 18, 44, 49
mycelium, 12, 38, 39, 43
mycophobia, 41
mycorrhizal relationships, 12, 73
myths
    cutting vs. plucking, 44
    gills, 43
    mushroom population, 43
    pick shaming, 44
    regeneration, 44
    touching, 41–42
    toxicity identification, 45

N

names
    common names, 10
    Latin binomials, 10, 136

O

oak trees, 153
old man of the woods *(Strobilomyces floccopus),* 102–103, 120
old man of the woods *(Strobilomyces strobilaceus),* 26
opportunities, 21
orange jelly fungus *(Dacrymeces palmatus),* 182
ornate-stalked bolete *(Retiboletus ornatipes),* 120
overexploitation, 48
oyster mushrooms *(Pleurotus genus),* 28, 43, 52, 111, 114–116, 128–130. *See also* late fall oysters *(Sarcomyxa serotinus)*

P

painted bolete (Suillus spraguei), 120
peach chanterelle *(Cantharellus persicinus),* 77
pear-shaped puffballs *(Apioperdon pyriforme),* 99
permission, 16–17
Phallus hadriani, 60
Phallus impudicus, 60
pheasant back *(Cerioporus squamosus),* 33, 55, 62–63, 64
photographs, 11, 14
pick shaming, 44
pinelitter gingertails *(Xeromphalina caucticinalis),* 174
poison ivy, 21
poison oak, 21
poison sumac, 21
polypore mushrooms, 33
Popcorn Shrimp of the Woods, 147
porcini *(Boletus edulis),* 120
pores, 33
puffballs
    brain puffball *(Calvatia craniiformis),* 98
    common puffball *(Lycoperdon perlatum),* 35, 92, 96–97
    giant puffball *(Calvatia gigantea),* 93–94
    Grilled Puffball Steaks, 95
    pear-shaped puffbals *(Apioperdon pyriforme),* 99
purple-gilled laccaria *(Laccaria ochropurpurea),* 158–159

R

Ramaria, 76, 123
raw mushrooms, 47
recipes
    Amber Jelly Roll Gummy Candies, 148
    Berkeley Jerky, 110
    Chanterelle Cream Sauce Linguine, 76
    Chicken of the Wood All-Purpose Seasoning Salt, 86
    Cream of Wild Mushroom Soup, 130
    Double Extraction Turkey Tail Tincture, 180
    Fried Morels, 59
    Grilled Puffball Steaks, 95
    Hedgehog Mushroom Bacon Penne Pasta, 176
    Hen of the Wood Flatbread, 133
    Lion's Mane "Crab" Cakes, 133
    Lobster Mushroom Duxelles, 154–155
    Popcorn Shrimp of the Woods, 147
    Reishi Tea, 109
    White and Black Trumpet Pizza, 84
    Wild Summer Mushroom Toast, 117
    Wild Yellowfoot Quiche, 173
regional field guides, 9
reishi mushrooms (*Ganoderma* sp.), 12, 101, 106–108, 157
Reishi Tea, 109
resinous polypore (*Ischnoderma resinosum*), 156–157
reticulum, 38
ringed honey mushrooms (*Armillaria mellea*), 142
ringless honey mushrooms (*Desarmillaria caespitosa*), 142
rings, 37
rooted agaric (*Hymenopellis furfuracea*), 118–119
rules, 16
Russell's bolete (*Aeroboletus russellii*), 163

S

sac fungi, 34
saffron milk cap (*Lactarius deliciosus*), 31
scales, 26, 38
scarlet waxcap (*Hygrocybe coccinea*), 81
shaggy mane (*Coprinus comatus*), 104–105
shaggy pholiota (*Pholiota squarrosoides*), 144
shaggy stalked bolete (*Austroboletus betula*), 38, 120, 162–163
shrimp of the woods (*Entoloma abortivum*), 145–147
size variations, 53
skirts, 37
slippery jack (*Suillus luteus*), 120
smooth chanterelle (*Cantharellus lateritius*), 72, 77
snakes, 19–20
snow fungus (*Tremella fuciformis*), 36, 150–151
soaking, 59
socks, 17
spiders, 20
spore prints, 15–16
spore-producing surfaces
    cup fungi, 34
    gills, 28–31
    morels, 34
    pores, 33
    puffballs, 35
    spore function, 28
    teeth, 33
    truffles, 34
    universal veil, 37
spring mushrooms
    deer mushroom (*Pluteus cervinus*), 15, 29, 68–69
    pheasant back (*Cerioporus squamosus*), 33, 55, 62–63, 64
    wine caps (*Stropharia rugosoannulata*), 64–65
    wood ear (*Auricularia angiospermarum*), 66–67, 149
stalks, 37
stems, 37, 38
stinkhorns, 35, 60
stipes, 37
striations, 26
strict-branch coral (*Ramaria stricta*), 124
subdistant gills, 32
summer mushrooms
    beefsteak polypore (*Fistulina americana*), 33, 100–101
    Berkeley's polypore (*Bondarzewia berkeleyi*), 89, 110–111
    black trumpets (*Craterellus fallax*), 82–84, 117, 130
    crown-tipped coral (*Artomyces pyxidatus*), 123–124
    Frost's bolete (*Exsudoporus frostii*), 120, 121–122
    golden chanterelle (*Cantharellus* sp.), 8, 73–74, 130
    hemlock reishi (*Ganoderma tsugae*), 107–109
    indigo milk cap (*Lactarius indigo*), 31, 112–113
    old man of the woods (*Streobilomyces floccopus*), 102–103, 120
    rooted agaric (*Hymenopellis furfuracea*), 118–119
    shaggy mane (*Coprinus comatus*), 104–105
    summer oyster (*Pleurotus pulmonarius*), 115, 116

T

teeth, 33
The Last of Us series, 41
ticks, 20
touching, 41–42
train wrecker (*Neolentinus lepideus*), 63
trash collection, 19, 21
tree guides, 12
trees, 12, 13, 14
trim and drop, 21
true gills, 30
truffles, 34

tubes, 33
turkey tail *(Trametes versicolor)*, 27, 33, 178–180
two-colored bolete *(Baorangia bicolor)*, 120

U

umbrella polypore (*Polyporus umbellatus*), 132
universal veil, 25, 37, 38

V

Verpa bohemica, 61
Virginia creeper, 21
volvas, 38

W

warts, 25
weather, 14
White and Black Trumpet Pizza, 84
white truffles, 34
Whitney Wisdom
    disappointment, 50
    discouragement, 48
    fun, 50
    gear simplicity, 49
    identification safety, 50
    littering, 48
    respect for nature, 48
    tolerance, 47
Wild Summer Mushroom Toast, 117
Wild Yellowfoot Quiche, 173
wine caps (*Stropharia rugosoannulata*), 64–65
winter mushrooms
    chaga (*Inonotus obliquus*), 168–169
    enoki *(Flammulina filiformis)*, 11, 16, 43, 170–171
    hedgehog mushroom (*Hydnum* sp.), 12, 33, 175–177
    late fall oysters *(Sarcomyxa serotinus)*, 166–167
    turkey tail *(Trametes versicolor)*, 27, 33, 178–180

witches' butter *(Tremella mesenterica)*, 36, 181–182
yellowfoot *(Craterellus tubaeformis)*, 12, 29, 172–174
witches' butter (*Tremella mesenterica*), 36, 181–182
wood blewit *(Collybia nuda)*, 113, 140–141, 142
wood ear *(Auricularia angiospermarum)*, 66–67, 149
wood ear *(Auricularia auricula-judae)*, 36
wood nettle, 21
woolly chanterelle *(Turbinellus floccosus)*, 75, 76
woolly milkcap *(Lactarius torminosus)*, 27
wrinkled cort *(Cortinarius corrugatus)*, 160, 161

Y

yellowfoot (*Craterellus tubaeformis*), 12, 29, 172–174
yellow morel *(Morchella americana)*, 57, 58

Z

zonations, 27

GO FORTH AND FORAGE!